Praise for *Hidden Games*

"This is one of those books that you pick up and then can't put down; next thing you know, you've been reading for hours and forgotten to go to bed (true story). Filled with quirky facts and perspective-shifting insights, *Hidden Games* explains some of human beings' most puzzling behavior with one of the most powerful theories ever devised: game theory."

> —Steve Stewart-Williams, author of
> *The Ape that Understood the Universe*

"In this fascinating book, Hoffman and Yoeli show how the tools of economics can be used to understand a wide range of real-world phenomena. The authors show time and again that many types of human behavior which seem inconsistent with consciously rational behavior can be understood once we realize that those same forces are operating below the surface. Indeed, this book shows the magic of what is hidden from view."

> —Kevin M. Murphy, MacArthur fellow and professor
> of economics at the Booth School of Business,
> University of Chicago

"*Hidden Games* is a great read. Hoffman and Yoeli show how widespread but puzzling behavior can result from self-interest, explaining subtle, important ideas in a snappy, accessible style."

> —Rob Boyd, author of *A Different Kind of Animal*

Hidden Games

The Surprising Power of Game Theory to Explain Irrational Human Behavior

Moshe Hoffman and Erez Yoeli

BASIC BOOKS
New York

Basic Books
Hachette Book Group
1290 Avenue of the Americas, New York, NY 10104
www.basicbooks.com

Printed in the United States of America

First Edition: April 2022

Published by Basic Books, an imprint of Perseus Books, LLC, a subsidiary of Hachette Book Group, Inc. The Basic Books name and logo is a trademark of the Hachette Book Group.

The Hachette Speakers Bureau provides a wide range of authors for speaking events. To find out more, go to www.hachettespeakersbureau.com or call (866) 376-6591.

The publisher is not responsible for websites (or their content) that are not owned by the publisher.

Figures designed by Jessica Weisman

Print book interior design by Trish Wilkinson

Library of Congress Cataloging-in-Publication Data

Names: Hoffman, Moshe, author. | Yoeli, Erez, author.
Title: Hidden games : the surprising power of game theory to explain irrational human behavior / Moshe Hoffman and Erez Yoeli.
Description: First edition. | New York : Basic Books, [2022] | Includes bibliographical references and index.
Identifiers: LCCN 2021045076 | ISBN 9781541619470 (hardcover) | ISBN 9781541619463 (ebook)
Subjects: LCSH: Game theory. | Human behavior.
Classification: LCC HB144 .H634 2022 | DDC 658.4/0353—dc23
LC record available at https://lccn.loc.gov/2021045076

ISBNs: 9781541619470 (hardcover), 9781541619463 (ebook)

LSC-C

Printing 1, 2022

CONTENTS

Hidden Games

CHAPTER 1

INTRODUCTION

FOR MANY OF US, IT MIGHT BE HARD TO IMAGINE THAT THE COVER OF *Sports Illustrated*, which we are used to seeing adorned by athletes and models sporting bats, balls, helmets, and bikinis, once featured the laughing, square-jawed face of Bobby Fischer. The 1972 cover, now a collectible on auction sites like eBay, celebrated Fischer's unprecedented twenty-game winning streak, not in baseball, basketball, or football—but in chess. Fischer's performance in the game, and his dominance over Soviet rivals, had catapulted him into the public eye like no other chess player before or since. Many of the game's greatest players considered him, and still do, to be the greatest of all time—the GOAT, to use the term usually applied to the likes of Michael Jordan, LeBron James, Simone Biles, Katie Ledecky, and Tom Brady.

How did Fischer, who grew up in Brooklyn in a cash-strapped, single-parent household, become so great? The answer is no different from the one that might be given for the likes of Jordan, James, Biles, Ledecky, and Brady: a touch of luck and a ton of practice, motivated by an obsessive passion. Frank Brady, Fischer's longtime biographer, reports that by the time he was nine, if Bobby wasn't playing chess then he was studying it, bent over his board or book in such a rapture that he wouldn't pause to turn on the apartment lights when it grew dark. To coax him into the bath, his mother would lay a cabinet door over the tub and place his chessboard on top of it. (Getting him to then leave the tub was another challenge.) He certainly couldn't be bothered with school, which he quit as soon as it was legal to do so.

Indeed, an obsessive passion is a key ingredient in the success of so many greats.

When Itzhak Perlman, the virtuoso violinist, first asked to play after hearing classical music on the radio at the age of three, he was denied admission to a local conservatory on the grounds that he was too small to hold a violin. So, the sickly boy, who soon contracted polio and today is still bound to crutches and a wheelchair, taught himself to play on a toy fiddle. By the age of ten, he was giving critically acclaimed recitals, and at nineteen, he appeared on *The Ed Sullivan Show*—for the second time—alongside the Rolling Stones.[1]

The mathematician Srinivasa Ramanujan was, in his short thirty-three years, so prolific that today an entire peer-reviewed journal is devoted to publishing results that derive or otherwise relate to those that he had claimed or proved. Like Fischer and Perlman, Ramanujan developed his passion at an early age, absorbing all he could from the college-aged lodgers who stayed at his mother's home and reading math textbooks cover to cover. One, a collection of five thousand theorems that is particularly credited with elevating his genius, would be impossibly tedious even to the most enthusiastic participant in a high school Olympiad. As an adult, his work engrossed him so fully that he neglected his wife and even his own health, dying from complications of dysentery that doctors now think could have been cured had Ramanujan bothered to take a break from work to receive care.

Or what about Marie Curie? She's still the only person to ever win two Nobel prizes in science. As a student in Paris, she was so engrossed in her studies that she often forgot to eat. She would remain so engrossed until her dying day, eschewing prizes and awards because they took her away from scientific pursuits. She even failed to acquire the funds from her (second) Nobel Prize out of what she termed "sheer laziness." (She finally picked up the award money during World War I so she could contribute it to the war effort.)

For Picasso, it wasn't chess, violin, math, or science but art, which he continued to produce obsessively for the entirety of his life. It is estimated that he produced over fifty thousand (!) works of art,[2] regularly reinventing himself, when any other successful artist likely would have rested on his laurels.

As mere mortals, the rest of us marvel at these obsessive passions. If only we could be drawn to shooting free throws as much as LeBron James, grinding through endgames as much as Fischer, or flipping through theorems as much as Ramanujan, we might be so much more successful! Each New Year, when we assemble our resolutions, we hope that we, too, magically develop the elusive passion that will transform our working hours from a chore into a labor of love. Yet, inevitably, by early February, when no such fire has been sparked, aimless Netflix and Instagram sessions overtake our aspirations. Why can't we be more like Bobby and Itzhak? How on Earth could Ramanujan have possibly been enthralled with a textbook of five thousand theorems?! Hand me the remote. Why were they, alongside Einstein and Picasso, anointed to receive the magic fire of passion, as though from Cupid casting arrows to a lucky few?

And why did they develop their *particular* passions—for basketball, chess, math, physics, or whatever? Why didn't Picasso, who was sympathetic to Catalan rebels and antifascists, devote his prodigious energies to war instead of art? Why didn't Einstein obsess over chess? Why would Fischer, who would sit still for hours as he poured over chess books and whose IQ was certainly no impediment, immediately grow restless as soon as a homework assignment of any kind was put in front of him (it inevitably went unfinished, to his mother's great consternation)?

In short, how does passion—this elusive maker of greatness—work?

There are entire fields devoted to judgment, decison-making, and positive psychology, and sections of bookstores devoted to self-help, so you'd think someone would have a ready answer to these relatively fundamental questions by now. How can we understand the decisions we make, or what makes us happy, if we don't understand what makes us passionate and gives our life meaning? Indeed, there are some things we know. We know, for instance, that passion goes hand in hand with a sense of meaning, purpose, and satisfaction and that it increases with praise or decreases if someone offers to pay us for our labors. But why? Why do passions work these ways? Do passions simply defy explanation?

No.

◆

EQUALLY PUZZLING: AESTHETICS.

Of course, some aspects of aesthetics have ready or well-understood explanations. We know why the wealthy and powerful pay for images of themselves, and why the church has long paid for art that teaches its myths and histories, which was especially useful back in the day when most parishioners were illiterate. We also know that art sometimes takes the things we already enjoy looking at, like symmetric faces, fecund women, or lush lakeside scenery, and exaggerates them. When it comes to music, some of it sounds like water or is otherwise soothing. And some of it keeps a steady beat, so that soldiers can march or townsfolk can dance. As for our food, we know it tastes better when it's more nutritionally dense. Hence bacon. And that in places where foodborne illnesses are a constant risk, we develop a taste for the spicy concoctions that help to inhibit the growth of bacteria.

But these leave unexplained so much of what's going on. What's with the complicated rhyming schemes of Renaissance bards like Shakespeare and modern ones like MF Doom, Chance the Rapper, and Eminem? Or with the highly tannic wines of Bordeaux's celebrated Left Bank? These aren't inherently more pleasant. Nor are they simply exaggerating what we already find pleasant. This is made clear by the fact that to the novice Shakespeare is incomprehensible and tannic wines taste too bitter and astringent (while Three Buck Chuck is perfectly quaffable). We don't mean to disparage these great works of art and culture. They're great, just not because they're inherently pleasant. So, what makes them great?

Also unexplained are the ubiquitous Easter eggs that artists of all genres litter in their artwork for critics and enthusiasts to discover by poring over the work for decades or sometimes centuries—while the rest of us to try to glean those hidden meanings from CliffsNotes or, for the exceptionally tenacious, from the ponderous writings of critics.

To find the Easter eggs, identify the best vintages, and understand Shakespeare, we can turn to art historians and critics. But to learn why—why we get excited by these things in the first place—we'll need some of the tools developed in this book.

◆

ALTRUISM IS ANOTHER DOMAIN WE'LL PUZZLE OVER. NOT JUST OVER why people are altruistic in the first place but also over the bizarre forms altruism takes.

For starters, it's clear that though we are quite caring and giving, we are not driven by impact and don't give in the most effective ways. We are moved to donate to GoFundMe campaigns for needy pets, instead of earmarking the funds for high-impact charities that can, by operating leanly and tackling some of humanity's most pressing problems, save a human life *for less than $5,000*.[3] When presented with a matching fund, we hardly respond, even though our money is going twice as far. When asked how much we'd donate toward safety nets that save migrating birds from being killed by wind turbines, we give the same response irrespective of whether we're told the nets would save two thousand birds or two hundred thousand.[4] We volunteer for Habitat for Humanity even though our airfare could have been better spent hiring local labor that is both more skilled and desperate for work. We care enough to turn off the lights when we leave the room but mindlessly leave the AC on, more than wiping out the gains from hitting the light switch.

We're not just ineffective, we're flat out ignorant. Most of us have, at best, a faint idea of what our donations are used for, and virtually none of us puts the same care into selecting a charity that we do into selecting a restaurant or vacation destination. We're just as bad when it comes to conserving energy and recycling. Did you know, for instance, that recycling metal has roughly nine times the impact of recycling paper or plastic? And that recycling paper or plastic is far more impactful than recycling glass? Don't believe us? Google it. But notice that this is probably the first time you've bothered to google it.

We're not just ignorant, we're *strategically* ignorant. We would never knowingly infect a sexual partner with an STI but are content with not getting tested, even if we know we are at high risk and testing is freely available at nearby clinics.

We don't just avoid the information; we also avoid the ask. We might give to Planned Parenthood if asked but pull out our phones and try to look really busy when we see the nonprofit's volunteers asking for donations on city sidewalks. And we'll, of course, always do a friend a favor but might avoid calling if we suspect said friend needs said favor.

There's more. Most of us have no qualms paying four dollars for a cortado rather than donating those funds to the poor. But we would never dream of taking four dollars from the poor to pay for a cortado. Why this distinction between action and inaction when the effect is the same? Why does altruism work this way?

THESE ARE THE KINDS OF QUESTIONS WE'LL TACKLE IN THIS BOOK. What tricks do cable news networks use to misinform? Why does motivated reasoning work the way it does? Internalized racism?

Why is modesty a virtue? Where does our sense of right come from? Why couldn't the Hatfields and McCoys bury their hatchets?

In short, we'll ask: Why are human preferences and ideologies the way they are? Why do they work the way they do?

People tend to respond to questions like these with proximate answers, for example: We love tannic wines because they are more interesting. We love crafts because it is satisfying to work on discrete projects with a finite timeline, where we can quickly see the end results, or we develop a passion for research because we like the freedom to engage in long, detailed explorations of a particular topic and really become experts in it. We give out of empathy for the recipient and do so ineffectively because empathy itself is a blunt tool that's not so sensitive to efficacy.

While such responses are often interesting, helpful, and valid, they aren't really answers, at least not in the sense we will be looking for in this book. Sure, tannic wines are more interesting, but what counts as interesting? And why do we even care if they're interesting? Sure, some people develop a passion only when they quickly see the results of their handiwork while others only get excited by longer, in-depth projects, but we're still left wondering why some are drawn in one direction while others are drawn in the opposite one, as well as why anyone develops any passion at all. Sure, empathy is a blunt tool, but why? Each of these answers raises at least as many questions as we started with!

We will, instead, attempt to give answers that are, in some sense, more ultimate. In doing so, the key tool we will be using is, of course, game theory.

Game theory is a mathematical tool kit designed to help us figure out how people, firms, countries, and so on will behave in interactive settings—when it matters not only what they do but also what others do. The tool kit has been successfully deployed to help firms design and bid in auctions (where how each bidder should bid depends on others' bids). It is also a cornerstone of federal antitrust regulation. At the Federal Trade Commission and US Department of Justice, armies of economists spend their days evaluating proposed mergers and acquisitions with the help of a game theory model called Cournot competition (which helps them predict how prices will change, taking into account that all firms in the market will react to what the merged firm does and vice versa). A few blocks away, at the US Department of State, game theory has influenced the thinking of generations of diplomats. For instance, the United States' cold war strategy of mutual destruction and nuclear brinksmanship was reinforced by the game-theoretic analyses of Thomas Schelling (which took into account that the number of nukes the US should make depended on the number that the USSR had and vice versa).

You might be thinking to yourself that this has absolutely nothing to do with the kinds of behaviors that the book opened with. People aren't even trying to optimize when they become passionate about playing chess, develop new art movements, or give to charity. They do these things based on intuition or feel or . . . it just kinda happens without them even realizing it at all. That sounds nothing like the cold-hearted calculus involved in boardroom and situation-room decision-making.

Moreover, you might also be thinking that game theory traditionally rests on a key assumption that's, well, let's say questionable: the assumption that people behave optimally. That we are rational. That we have all the relevant information and use it as a computer might to maximize its benefits—doing complex calculations in the process. Maybe this assumption is decent for the crew in the boardroom, strategizing over their radio spectrum bid, but for the rest of us going about our day-to-day lives? There have been not one but two Nobel Prizes in economics for emphatically knocking that assumption down (Daniel Kahneman's in 2002 and Richard Thaler's in 2017).[5] Even some of our motivating puzzles— willingly dying for a cause, giving to ineffective charities when effective ones stand at the ready—seem to be strong evidence in Danny and Dick's favor.

We are going to use these two arguments to cancel each other out. Yes, people are quite often quite bad at optimizing *when they are relying on their conscious minds to do the optimization.* But when they are not consciously optimizing, and it is learning and evolution doing the optimization—as we will argue is often the case for tastes and beliefs—things start to look a lot more promising.

When it comes to evolution, the logic is likely already familiar. People's tastes evolved to motivate us to act in ways that benefit us. We evolved a taste for fatty, salty, and sweet foods because that motivated us to seek out foods high in fat, salt, and calories in an environment where these were rare. We evolved an attraction to symmetrical faces, chiseled jaws, and broad hips because this motivated us to seek partners who were more healthy, successful, and fertile.[6]

But it's not like rap fans evolved from caveman ancestors who sat around the fire trading rhymes while modern art fans' ancestors devoted their leisure time to abstract cave painting (*"Ceci n'est pas une mammouth laineux"*). Most of the tastes and beliefs we're interested in aren't biologically ingrained in us. They're learned. So, in the next chapter, we'll make the same argument for learning that we just made for biological evolution and show that learning (aka *cultural* evolution) does the job just as well (and quite a bit faster). We'll see how cultural practices end up being highly tuned to our environment and needs: how, for instance, the igloo was tweaked and tuned over generations until it could keep the Inuit warm in the icy tundra or how traditional methods of preparing corn eked out extra nutritional value from this nutritionally sparse food staple—and without anyone consciously thinking about thermal dynamics or chemistry. We'll also see how the spices people come to savor reflect their culture's need to combat foodborne illness and how superstitions and taboos surrounding foods reduce the risk of dangerous illnesses during pregnancy.

After the chapter on learning, we will cover a few distinctions (primary versus secondary rewards, ultimate versus proximate, emic versus etic) that will help us interpret game theory when it is hidden—working its magic on our beliefs and preferences behind the scenes, with the help of evolutionary and learning processes.

After that, we will finally get to some game theory. But it won't yet be focused on humans. Instead, we will have a chapter devoted to animal sex ratios—the ratio of males to females in any given species—which is a well-known application of game theory from biology. This chapter will introduce some of the key concepts in game theory and show off just how powerful it can be. It will also help us to see how a game is interpreted and applied when evolution is doing the optimizing.

Then, we're off to the races. From that point out, each chapter will present a handful of seemingly irrational human behaviors and a hidden game or two that will help us uncover the underlying rationale behind these seemingly irrational behaviors.

That's the plan. Shall we get to it?

CHAPTER 2

LEARNING

IN THIS CHAPTER, WE WILL LAY OUT JUST HOW POWERFUL LEARNING processes are at getting us to behave optimally. We will see that the power of learning is not limited to cases where we have any awareness of what is being optimized. Nor is the power of learning limited to our behaviors; often our beliefs and preferences are sculpted as well.

Why talk about any of this? Because it lays the groundwork for using game theory even when people aren't rational and even when we are trying to explain their puzzling preferences and beliefs.

REINFORCEMENT LEARNING

In video footage from the 1950s that can still be found floating about on YouTube,[1] B. F. Skinner stands before some green and white lab equipment in a dark tie and a white button-down, with the sleeves pulled down all the way to his wrists. Speaking into the microphone in a pleasing mid-Atlantic accent that was made for television, Skinner narrates as he trains a pigeon to spin about in a counterclockwise direction. Skinner's strategy is simple. Each time the pigeon turns to her left, Skinner opens a trough and gives the pigeon a treat of a few grains, but if the pigeon stands in place or turns to the right, Skinner leaves the trough closed.

The camera zooms in on the pigeon, who, at first, just bobs back and forth. Eventually, randomly, she shuffles half-heartedly toward the left. Click. The trough opens, and the pigeon quickly nabs the snack. The trough closes, and the pigeon looks about. Again, she bobs a bit, and shuffles about somewhat aimlessly.

"I'm waiting for it to turn counterclockwise now," Skinner intones, "then I reinforce that movement." Soon, the pigeon shuffles toward the left and—click—Skinner opens the trough.

The pigeon has caught on now. As soon as the trough shuts, she shuffles to the left expectantly.

"You see, the effect is instantaneous," Skinner says with some pride. The pigeon pauses, and Skinner adds, "I'm waiting for a more pronounced movement than that. It's got to be more than that." The pigeon shuffles a bit more. Click. The pigeon hurries back to grab her prize.

The trough closes. The pigeon immediately turns to the left and, with barely a pause, makes a full circle.

"There we go, all the way around," Skinner announces flatly. The whole training session has taken less than a minute.

Skinner's video is a stark illustration of reinforcement learning, one of the key processes that shapes animal and human behavior. The core essence of reinforcement learning is straightforward: when a behavior (like shuffling counterclockwise) leads to a reward (like a morsel of food) then it is reinforced and is more likely to be repeated.

Reinforcement learning is ubiquitous. Anyone with a pet has employed reinforcement learning. Dogs learn to sit and stay because we give them treats when they do what we say. Cats learn to stop scratching the couch when we spray them with a water bottle. If you don't have a pet yourself, an adorable way to see reinforcement learning in action is to search YouTube for videos like "How to train a pig" or "Teaching pigs to sit." Then again, you've already seen reinforcement learning in action since you—and everyone around you—learns via reinforcement. Kids learn to use the potty in pursuit of gummy bears, and they learn addition and sub-traction in pursuit of gold stars. Kids and adults both learn when they've told a good joke based on whether they're rewarded with a hearty laugh or nervous silence. And whether an outfit should be worn again or taken to the thrift shop based on whether it is complimented.

Reinforcement learning is powerful. Kids don't just learn simple arithmetic through reinforcement, they also learn long division, al-gebra, geometry, and precalculus. YouTube's pigs have been trained to complete obstacle courses, to score goals, or to ring a bell when they need to be let out to go potty, all through reinforcement. Skin-ner, famously, taught his pigeons to play ping-pong; in the mid-1990s, a team of Japanese researchers one-upped him by training their pigeons to accurately distinguish Picassos from Monets.

Although reinforcement learning can help animals end up on YouTube, this is obviously not what it's there for. It's there so that they learn functional behaviors that are critical for survival in a changing environment. Through reinforcement, animals learn where to find food, shelter, and mates, how to avoid predators, and which foods are poisonous or nutritious.[2]

SOCIAL LEARNING

The Yasawa Islands are a beautiful, remote volcanic archipelago in the Pacific. Although technically part of Fiji, for the better part of the twentieth century, the islands were ruled autonomously, and their king did not allow tourism on the islands. When the anthro-pologist Joe Henrich and his PhD student James Broesch visited Yasawa in the mid-2000s, here's what they found:[3]

Economically, Yasawans rely primarily on horticulture, fishing and littoral gathering. Fishing is the most important source of protein, and spear-fishing is the most productive form of fishing for those with sufficient skill. People also fish with lines and nets.

Yams and cassava provide the caloric staples, although yams are preferred, traditional and necessary for ceremony life. Men compete informally to grow the largest yams. Political units are composed of interrelated clans called Yavusa, which are governed by a council of elders and a hereditary chief. Social life is organized by a complex web of kinship relations and obligations. At the time of the study there were no cars, TVs, markets or public utilities in these villages.

In Yasawa, Henrich and Broesch set out to understand how the Yasawans learned the skills necessary for survival: to fish, grow yams and cassavas, and use medicinal herbs. They surveyed the local population, asking questions like: Who would you go to for advice if you had a question about fish or fishing? Who would you seek advice from about planting or growing yams? Who would you ask about using which plants to use for medicine? They also asked questions like: Who are the best line fishers? Who are the best yam growers? Who knows the most about medicinal plants? They jotted down the names people mentioned and collected data about those individuals. How old were they? What was their sex? Were they in the same village? The same household? Their findings are straightforward. By far, Yasawans tended to ask advice from the individuals who were the best at fishing, growing yams, and using medicinal plants.

The Yasawans' answers to Henrich and Broesch's questions illustrate two important points. First, we don't just learn from our own experience via reinforcement, we also learn from others by imitating them (sometimes also by explicit instruction). If you want to learn to fish, farm, or use medicinal plants, you don't need to risk starvation and sickness; you can ask for advice or copy someone who seems to know what they are doing. If you want to figure out whether you should wear a particular outfit, wearing it and (proverbially) fishing for compliments, and adjusting your wardrobe based on the strength of such compliments, is not your only option. You can also look around and see what others are wearing.

Second, when we learn from others, we don't do it randomly. Our learning is biased in a number of ways that make us prone to learn from those in the know. We imitate more those who are successful, prestigious, older (or of our age group, if what we are learning is age specific), reasonable, and so on. In the Yasawans case, they sought advice from those who were best at fishing, farming, and medicating with plants. In the case of our wardrobes, we attend most to style icons like Michelle Obama and George Clooney.

In one experiment, fourteen-month-olds were more likely to imitate an adult who switched on a light using a nontraditional method (pushing the switch with his forehead) if the adult had behaved competently in the past (putting his shoes on his feet) than if he had behaved incompetently (acting confused and putting his shoes on his hands).[4] In another study, preschoolers learned new words by imitating but were more likely to imitate an adult than another child; however, preschoolers didn't take just age into account when learning new words, they also prioritized reliability. They were more likely to imitate someone who had reliably used words that they already knew, whether child or adult, than someone who used them incorrectly and only used age to break a tie: if a child had been reliable but an adult unreliable, they imitated the child, whereas if both had been unreliable, they imitated the adult.[5]

Children also recognize that adults know better about some things (like which foods are nutritious) while kids know better about other things (like which toys are fun)[6] and that there are situations where it makes more sense not to imitate others, including adults. In the original toddler-light-forehead experiment, researchers included a version in which they hamstrung the adults by tying their hands but left the toddlers' hands untied. The toddlers didn't bother imitating adults whose hands were bound, seemingly concluding that the adults were only turning on the lights with their foreheads because they were hamstrung and, thus, there was no benefit using their foreheads when they themselves had no such hamstringing.[7]

One of the premises on which the analyses in this book rest is that learning, regardless of whether it is from one's own experience via reinforcement or from others' via imitation and instruction, leads us to do what is good for us, at least on average, much of the time. The fact that learning is so sophisticated helps make us confident in this premise.[8]

Next up are two cases that illustrate this point, while also highlighting another one: that we're often completely unaware of why we do what we do or the role that learning has played.

THE IGLOO IS A REMARKABLE ACHIEVEMENT OF HUMAN INGENUITY. Built using only locally available materials (packed snow and the occasional seal skin; there is no wood, no brick, no stone, and no clay in the Arctic), an igloo's interior can warm by 60 degrees from just a small oil lamp, even as Arctic winds rage outside.

This is no mean feat. Indeed, just building a structure out of snow that won't collapse is difficult enough, as many readers who attempted to build igloos as kids probably know from experience. To keep the igloo from collapsing and make it strong enough to withstand those powerful Arctic winds, the Inuit assemble blocks of snow into a very strong arch called an inverted catenary (a catenary is the shape a piece of string or a necklace will form if you hold it up by two fingers). Since its discovery in ancient times, the inverted catenary has been used in the construction of monuments, buildings, and bridges.

There are several key factors that help make an igloo as warm as it is. The packed snow from which the igloo is constructed is, somewhat counterintuitively, an excellent insulator thanks to the millions of air bubbles trapped inside. Ice is not as good an insulator, so it is used sparingly, mostly in the construction of windows, which allow light in. The floor of the igloo is terraced: the Inuit sleep on the highest terrace, cook and perform other active tasks on the middle terrace, and enter through the lowest terrace, which is dug out below the surface of the surrounding snow pack. This ensures that heat emitted from lamps, cooking fires, and the Inuit's own bodies rises up past the entranceway and stays trapped inside. The entranceway itself is placed at a 90-degree angle to the prevailing winds and is often shaped with a right angle to further help in keeping the frigid winds at bay. All these details, and more, have been perfected over the generations.

It is just astounding that the Inuit figured all this out, but crucially, they didn't do it by calculating the load-bearing potential of an inverted catenary, the insulation value of packed snow, or the pressure differentials generated by different entranceways. Nor was it possible for a single individual to arrive at the design through trial and error without perishing. Presumably, one year an Inuit family dug out their entranceway. Their igloo was warmer than their last one and warmer than neighboring igloos. The next year, the family made sure to build their igloo the same way (reinforcement learning) and some neighboring entranceways were similarly dug out (imitation and instruction). And so on, until everyone built their entranceway in this way. The walls, windows, and vents were perfected via the same process. Over many generations, the Inuit learned their way to a perfect igloo, and there was no need for them to be aware that they were learning or why the igloo was shaped the way it was.[9]

Let's leave the icy Arctic and head to the warmer climes of Central and South America. There, dating all the way back to Mayan and Incan times, it has been the custom to soak corn in an alkaline solution prior to eating it. Traditionally, this was often done by tossing some burnt seashells or a bit of wood ash into the water in which the corn is boiled and then soaked. Once so treated, the corn kernels—now called hominy—turn a pale yellow and have a soft, gummy texture. They can be added to dishes whole or easily ground down into flour.

Treating corn with an alkaline solution in this way is called nixtamalization, and while it might initially seem like an odd thing to do, it is essential for releasing vitamin B-3, also known as niacin, in the corn, which is otherwise indigestible; without nixtamalization, people who depend on corn as their primary dietary staple will eventually develop nutrient deficiencies like pellagra, which causes dementia, diarrhea, and dermatitis. Of course, nixtamalization is centuries old—much older than the discovery of vitamins. So again, we see that a culture has adopted a useful practice without the aid of modern science. One day, a cook spilled some ash into the pot, presumably by accident. The folks who ate the corn must've felt and looked better. Or maybe they just appreciated that it was easy to mill and noticed the effects on their health a bit later. Maybe they didn't notice the health effects at all, but they were sick less often, and this meant they were generally more successful and therefore more likely to be imitated or sought out for advice. Regardless, reinforcement learning, imitation, and instruction had plenty of opportunity to work their magic, and there was no need for anyone to notice them at work or understand what they'd accomplished.

Just because nobody has a clue that there's a function doesn't mean there isn't one. This was a lesson that Europeans learned the hard way when they first encountered corn. Having failed to get a satisfactory answer as to why the natives nixtamalized the corn ("That's just how we've always done it"), the Europeans enthusiastically adopted corn as a food staple but skipped the silly-seeming nixtamalization. After all, they had powerful mills that could break down the corn into a flour without first softening it. Mass outbreaks of pellagra followed in Northern Italy, France, and the American South. In the first half of the twentieth century, some three million Americans suffered from pellagra; one hundred thousand died.

Humans are uniquely adapted to learn and imitate complex behaviors whose function is difficult to ascertain. One way we do this is by *over*imitating—imitating behaviors that *seem* entirely unnecessary (unless you are an overconfident, ethnocentric

European colonialist?). Classic studies showing overimitation look something like the following: children and chimpanzees are shown by an adult human how to open a box with a treat inside. The adult human has been instructed to add a few extra, unnecessary steps: maybe to tap the top of the box three times or touch his nose. The children imitate every move the adult makes, even the unnecessary ones. The chimps are wiser and leave out the silly steps. Wiser, perhaps, but more liable to make the same mistake Europeans did when it comes to things like nixtamalization.

◆

CONTINUING ON OUR TRAVELS, WE RETURN TO FIJI, WHERE WE WILL see a nice illustration of how learning doesn't shape just behaviors but also beliefs.

When Fijian women become pregnant and also while they breastfeed, they strictly abide by food taboos that prohibit eating certain fish such as rock cod, shark, barracuda, and moray eel— fish that are regularly eaten at other times. (Typically, the Fijians get their protein from land meat, as well as shellfish, octopus, and porcupine fish.)

Unbeknownst to the Fijians, the very foods they have learned to avoid are those that carry the dangerous toxin ciguatera. Too much ciguatera can make us sick, causing pain in our hands and feet, or very bad diarrhea. These symptoms can sometimes last for months. Ciguatera poisoning is especially dangerous during pregnancy and breastfeeding because women are more susceptible to the toxin during this period and because it can harm fetuses and breastfeeding infants.

Ciguatera is a chemical produced by algae. It accumulates in fish that eat the algae, accumulates in higher concentrations in the fish that eat those fish, and in even higher concentrations in the fish that eat those fish. Rock cod, shark, barracuda, and moray eel are all relatively high on the food chain (moray eels have even been known to attack sharks), so they carry with them the greatest risk of ciguatera poisoning. Octopus and porcupine fish are further down the food chain, and shellfish even further down. So, the risk of ciguatera poisoning from eating them is lower. The risk from land meat and vegetables is practically nonexistent. Fijian food taboos are, therefore, an effective nudge away from higher-risk foods toward lower-risk ones and thus toward healthier fetuses and babies.

Once again, it's remarkable that Fijians arrived at these highly functional food taboos via learning, without any awareness of their function. When asked where they learned these taboos, the vast majority answer that they learned them from their mother, grandmother, mother-in-law, an elder, a wise woman, and/or an aunt; fewer than 10 percent say they learned them from a doctor, and none of them have heard of ciguatera.

Another thing that's remarkable: it's not just Fijian's actions that are being shaped by learning and imitation, it's their very beliefs. When asked why they follow food taboos, most women answer that it can lead babies to have negative health consequences like "rough skin" or "smelly joints." It doesn't matter that these beliefs are wrong; they still work. (Don't be surprised when we point out the same thing about some of your beliefs.)

We mentioned earlier that one premise of this book is that learning leads to functional behaviors (or tends to, on average). Fijians' wrong but functional beliefs highlight a second, closely related premise: that the corresponding beliefs often come along for the ride. A great way for learning to get us to behave in a certain way (for instance, to avoid certain foods) is to shape our beliefs—to get us to believe that these foods will lead our baby to have rough skin—in a way that motivates us to act.

IT'S NOT JUST BELIEFS THAT CAN BE SHAPED BY LEARNING OR IMITA-tion; tastes can, too.

Visit Russ & Daughters, the iconic Jewish deli that has served Manhattan's Lower East Side for over one hundred years, and you are immediately surrounded on all sides by refrigerator cases overflowing with gleaming lox and whitefish and tub after tub of flavored cream cheeses. On the walls, there are baskets full of bagels and shelves heaped high with loaves of pumpernickel and rye breads, alongside colorful cans of imported caviar. This is a temple to the beloved cuisine of the hundreds of thousands of poor Eastern European Jews who passed from Ellis Island to the Lower East Side in the late 1800s and early 1900s. It is a temple—to salt. Salt is the primary seasoning you'll find at Russ & Daughters. You might, if you look hard, find a bit of pepper here and there. And, of course, the obligatory "everything" seasoning on some of the bagels, resting in their baskets. There's also some dill, if you count that. And some onion, now and again. But, mostly, the delicacies at Russ & Daughters are seasoned just with salt.

Walk just a few blocks to Manhattan's Curry Row on East Sixth Street, and things couldn't be more different. At Jewel of India—one of the many restaurants where barkers will rush you as you walk down the sidewalk—you'll find fragrant balti curries and searing vindaloos, each made with over a dozen spices and enormous quantities at that. Ask the waiter what he thinks of bagels, lox, cream cheese, and caviar, and he'll tell you, with a sideways nod, "It's all right." Ask him again, and he'll admit, "Kind of bland. Very bland."

How come Indians developed a taste for very spicy cuisine? Why didn't Eastern Europe's Ashkenazi Jews do the same? In 1998, Jennifer Billing and Paul Sherman proposed that the answer lay in spices' ability to inhibit and kill bacteria that cause food to spoil. This ability, they figured, was most useful in hot climates, so this is where we'd expect to find people who have learned to love spice.[10]

To test their theory, Billing and Sherman began by documenting that spices do indeed have the ability to inhibit and kill bacteria. They did this by considering the thirty bacteria that most often cause food poisoning and then combining dozens and dozens of studies that tested whether the presence of a spice, or its active ingredients, slowed the growth of—or outright killed—one of these bacteria. Some spices, like allspice and oregano, and the root vegetables garlic and onion inhibit the growth of all thirty types. Many, like bay leaf, mint, coriander, and nutmeg, inhibit half to three-quarters of the bacteria. Some, like black pepper, and the citric fruits lemon and lime only inhibit a small number of bacteria on their own but amplify the effect of other spices (black pepper by increasing the bioavailability of other spices' active ingredients, thus increasing the rate at which they are absorbed by bacteria; lemon and lime by breaking down the bacteria's cell walls, making the bacteria more susceptible to the active ingredients in spices).

Then, Billing and Sherman compiled hundreds of recipes from dozens of cultures around the world and indexed the types of spices and quantities used. Sure enough: "As mean annual temperature . . . increases, the proportion of recipes containing spices, number of spices per recipe, total number of spices used, and use of the most potent antibacterial spices all increase." They then ruled out one alternative theory after another. It is natural to ask whether spices are simply used in hotter places because that's where they grow. But, as the existence of a bustling and ancient spice trade suggests, spices are used in many more places than they're grown. Allspice is a nice example: it is used in ten times as many countries as it is grown. Moreover, it turns out spices aren't even, on average, more commonly grown in hotter climates. In many hot places, few spices grow at all, but they're still widely used. Wait, maybe spices can help people stay cool by causing them to sweat? Indeed, the capsaicin in chili peppers has this effect. But most spices, like oregano, mint, and cinnamon, do not. Maybe spices contain micronutrients that are somehow more useful in hotter places? Nope. Locally available vegetables and meat contain much higher concentrations of critical micronutrients.

So, back in their home countries, Indian mothers gradually trained their babies to tolerate and then love spicy food (by mixing

adult food with yogurt and gradually reducing the amount of yogurt), while Ashkenazi mothers stuck to salt. When they came to
America, they brought their cuisines with them.

As with the Fijians and Inuit, these Indian and Ashkenazi mothers need not have known why they ate the way they did. Most
would have no idea and might respond: "That's just how we've
always done it," or "It tastes good this way." Nor would they
have been aware of the learning processes that worked their magic
over many generations, to shape their behaviors—by shaping their
tastes.

Why might learning shape beliefs and tastes, rather than just
shaping the desired behaviors directly? One possibility is the one
we raised earlier: it's simply an effective way to get us to act. If you
like spicy food, you'll eat spicy food. Another possibility, though,
is that internalization ensures the functional behavior is taken even
when its function is not understood. Like overimitation. If Indians
like the taste of black pepper in their curry, they won't be tempted
to remove it, even if they don't realize that it's there because it
amplifies the effect of cumin, coriander, turmeric, and capsicum.
If Europeans had had a chance to develop a taste for nixtamalized
corn, they'd have avoided pellagra.

LAGS AND SPILLOVERS

Remember our waiter at the Jewel of India, who sheepishly admitted to thinking bagels and lox are kinda bland? He's not alone.
Visit an Indian American family during Thanksgiving, for instance, and you might find that the turkey has been lightly curried
or marinated overnight in yogurt and tandoori spices before being
roasted whole. The cranberry sauce might look from afar like any
other but reveal itself to have a touch of heat from green chilis and
may be spotted with earthy, toasted cumin seeds. The gravy might
have the golden yellow glow and unmistakable fragrance of saffron and perhaps a bit more black pepper than usual. Nor is it just

Indian Americans. Visit a Mexican American household, and you might find the turkey served alongside pozole and other traditional dishes from south of the border. One Mexican American friend recounts the time her aunt was charged with making green bean casserole. Her aunt looked up the recipe, decided it made no sense, and showed up with green beans and chili, instead. The entire family was thrilled.

Of course, neither the lox at Russ & Daughters nor the Thanksgiving turkey and green beans are at serious risk of having spoiled. The American food supply benefits from abundant refrigeration, as well as industrial food safety practices and regulations that minimize such concerns. Rather, immigrants from warm locales skip the lox or spice their turkey and fixings because they like the spice. They have, figuratively and literally, developed a taste for it—a taste that developed in a culture that needed to protect against foodborne illness but that persists even though they now live in a culture with no such need.

We will call such effects lags. Lags are like vestigial traits—human tails or the whale's hand. They once had a purpose, but that purpose is no longer relevant. Lags occur because, like evolution, learning isn't instantaneous. Just as it takes time for an individual to learn to like spicy food and for a culture's cuisine to become spicy, it takes time for these things to be unlearned, or, as animal behaviorists would say, for extinction to occur.[11]

Sometimes, we also run into situations where a belief or taste still has a relevant purpose in one context, but it persists even when taken out of this context. We call such effects spillovers. When your dog humps your leg, that's a spillover. After all, humping some things has a very useful purpose, but a leg? Not so much. Spillovers occur because learning, like evolution, involves generalization—applying what we've learned in one context to other similar contexts—and while we're pretty good at generalizing, we're not perfect at it. Sometimes, we overgeneralize, and this is especially true when learning shapes our internalized beliefs and tastes.

Spillovers will be particularly helpful for understanding the results of psychology and economics laboratory experiments (that are, by design, performed in highly controlled laboratory settings). The laboratory environment is extremely useful: by tricking our tastes and beliefs into spilling over from day-to-day life into our experimental settings, we can document their many, fascinating quirks. (Nowadays, the word *laboratory* is in fact a bit of a misnomer. In many studies, the "laboratory" is Amazon Mechanical Turk, where people from far and wide can make a few extra bucks by logging in and anonymously answering surveys in their pajamas.)

Consider the following classic experiment. Subjects are partnered up in pairs to play the ultimatum game. One subject is given a few dollars—say ten dollars—and chooses how much to share with her partner. Her partner then chooses whether to accept the offer. If he accepts, they each keep their share. If he rejects, neither gets any money. Then, the subjects go their separate ways, never to meet again (literally—remember, these are strangers, on the internet, who have no way of identifying each other or even of communicating).

In the ultimatum game, it is very common to see fifty-fifty splits. Pretty intuitive. It is also common to see that when the first subject offers too little, say one or two dollars, her partner rejects the offer. Again, intuitive. And it makes sense: after all, our sense of justice is designed to prevent people from walking all over us, and when they do, we must strike back (see Chapter 14). The experiment is working beautifully: our sense of justice is nicely spilling over to this silly little game.

But wait. The experimenter has carefully designed the experiment so that it is anonymous. She has even made it so that she herself cannot identify the subjects. Nobody—not even the experimenter—will find out if certain subjects are pushovers and accept low offers. What's the use in being outraged? Subjects should, if they are behaving rationally, bite the bullet and accept the low offer, which is better than nothing. So, does this mean people's sense of justice does not have the function we claimed—of keeping others from trampling them? We don't think so. That would be like concluding that because dogs sometimes hump people's legs, their sex drive didn't evolve for reproduction. Rather, the correct interpretation, we think, is just that our sense of justice isn't perfectly tuned to the specifics of every situation—not surprising given how unusual the fully anonymous laboratory setting is and how learning processes lag and spillover.

NEXT, WE'LL TAKE ONE MORE CHAPTER-LENGTH DETOUR TO DISCUSS A few key concepts that will help us better understand how the game theory will be used and interpreted.

CHAPTER 3

THREE USEFUL DISTINCTIONS

IN THIS CHAPTER, WE'LL DESCRIBE THREE USEFUL DISTINCTIONS THAT will clarify exactly what kinds of explanations our games tend to provide. These distinctions are motivated by the important role that learning (and evolution, too) is playing in our analyses.

PRIMARY VERSUS SECONDARY REWARDS

When people discuss learning in animals, they typically distinguish between primary rewards and secondary ones. Food is an example of a primary reward. It's something every animal has evolved to like and something that's nearly impossible to get an animal to stop liking (which makes sense because if animals could easily learn to stop liking food, they'd be at risk of starving). As B. F. Skinner showed us, primary rewards like food are the core tools at the disposal of an animal trainer.

Of course, animal trainers have other tools at their disposal, like the tone of their voice (good boy!), pets and belly rubs, and clickers and whistles. The trouble is, many animals don't start out responding to these things. The trainer has to first associate vocal commands, belly rubs, and whistles with food. After enough association, the animal eventually learns to like hearing "good boy" or getting a belly rub independently of the treat, and then these, too, can be used to reinforce desired behaviors. Vocal commands, belly rubs, and whistles are examples of secondary rewards. These are things the animal didn't evolve to like but can learn to like—or to stop liking, if the secondary reward is not associated with a primary reward for long enough (this is called extinction). Animal trainers are careful to think of primary rewards and secondary rewards as distinct and to pair secondary rewards with primary rewards to avoid extinction.

The distinction between primary and secondary rewards can be applied to humans, too. There are some things that humans evolved to like and can't easily learn to stop liking. Food, obviously, is one. Good health is another. Comfort and safety. Time and effort. Trust, or at least the resources and relationships that come with it. Prestige, power, and sex are also contenders.

Meanwhile, there are plenty of other things that humans genuinely like and enthusiastically pursue, like stamp collections, abstract art, job titles, and the taste of a particular spice blend from their youth. But no human evolved to like these. Rather, they learned to do so because they were associated with primary rewards. Like belly rubs and whistles, these are secondary rewards—things we genuinely like but also can, over time, stop liking.

In our analyses, we have to be careful to keep the following in mind. We are trying to understand the tastes that people have—things like an obsession with stamps. We're doing that by uncovering the primary rewards that shaped those tastes. Game theory is going to help us with this, but only if we are careful to keep in mind that the payoffs in these games are not the preferences we have come to have—secondary rewards—but the primary rewards we started with. That's why we're emphasizing this distinction.

Before we move on, here are some pointers for identifying primary rewards. To tell if something is a primary reward, we ask ourselves the following questions.

- **Is it universally liked?** Food and sex are pretty universally liked. Spice, very dark chocolate, and Picasso's *Guernica*, less so. If it's not liked by everyone, those who like it probably had to learn to like it, and it ain't primary.

- **Is it something we had to learn to like?** Infants must be (slowly!) trained to like spicy dishes.[1] That's another hint that our taste for spicy food is not a primary reward. No one has to learn to like eating or enjoy good health, making these better candidates for primary rewards.

- **Can we unlearn it?** Once we like spicy food, it might be hard to unlearn the taste, yet even die-hard spice fans find that when they visit home after many years abroad, the food is spicier than they remember. What about a passion for chess or go? Will we learn to stop liking such games if people start telling us that we aren't actually good at them, or that they're a waste of our time? On the other hand, it is very hard to learn to stop liking sex, fatty foods, or a good reputation, another hint that these are primary.

- **Is it evolutionarily sensible?** Evolution would only instill in us primary rewards that were pretty consistently tied to survival and reproduction. Food, shelter, and social ties have always been essential for survival and reproduction. Makes sense why these tastes would be hardwired. But what about a desire to spread joy throughout the world or to advance racial equality? Why would evolution imbue us with a desire to devote resources to helping people in faraway lands? It wouldn't, because that often wouldn't be of great help to us. But it could certainly imbue us with the ability to learn such things under the right conditions.

- **Is it suspiciously flexible?** Our desire to help those in need is suspiciously susceptible to whether we are being observed helping, whether helping those people is normalized, whether there is plausible deniability for not helping them, and so on. All clues that helping is not itself a primary reward but more of a means toward some other ends (we will discuss these ends in Chapters 7 and 8).

Now that we have a sense of what primary rewards are and how to identify them, it's maybe worth emphasizing some things they are not.

- **Fitness.** Even though primary rewards must have been tied to biological fitness in our evolutionary past, they don't necessarily correspond to fitness today. We evolved to pursue (and to have learning processes that help us pursue) sex, status, and resources, in their own right, even when they no longer lead to survival and reproduction. We pursue sex even when birth control is involved, and status even if it won't help us obtain more mates (because we are no longer sexually active or are faithfully devoted to one mate). We also pursue wealth, even though, nowadays, wealthier people have fewer children. Primary rewards evolved because of their *historic* association with fitness, but they are not the same thing as fitness.

- **Conscious goals.** Primary rewards are also not the same as the goals we consciously pursue, which include collecting stamps, becoming a chess master, learning art history in depth, stocking an impressive wine cellar, and making the world a better place. We think it's better to think of most such things as secondary rewards—things we didn't start off liking but only learned to like because of their association with primary rewards. Passions, for instance, are not primary rewards but secondary. Of course, sometimes we do consciously pursue our primary rewards. When we are hungry, we might consciously seek food, and when we are, uh—well, anyway—we might sometimes consciously seek sexual partners, too. So, just because we consciously pursue something doesn't mean it can't be a primary reward. It just doesn't always mean that.

- **Psychological rewards.** Like conscious goals, the feelings we get when we give to charity (warm and fuzzy!) or when our beliefs and actions are inconsistent (dissonance!) are often learned and highly context dependent. In this book, we will often use primary rewards to explain where such psychological payoffs come from and why they have the odd features they do, but in so doing, we will never use psychological payoffs themselves as part of our calculus. This is a bit different from familiar social psychology and behavioral economics texts (like, say, Dan Ariely's *Predictably Irrational*). We'll pilfer a host of fascinating puzzles from such researchers, but for us, the puzzles won't be part of the explanation—they'll be the things that need explaining.

- **Financial incentives.** Lastly, primary rewards are not the same as money, which is what most people mean when they talk about incentives. That's not to say financial incentives aren't important and powerful. If you've read Steven Levitt and Stephen Dubner's *Freakonomics*, you've seen how once you sleuth around and uncover the financial incentives at play, they can explain why sumo wrestlers take falls for an opponent who needs the win more than they do (the opponent offers to share the winnings), teachers inflate their students' test scores (to avoid pay deductions), and real-estate agents hurry to sell others' houses but have all the time in the world when it comes to selling their own (higher offers matter much more if they are the seller and keep the entire proceeds rather than just a small commission). Financial incentives don't, however, include a swath of the incentives we will be interested in, many of which are purely social—like trust, prestige, or romantic partners. Such primary rewards will also often operate outside our conscious awareness as they shape our tastes and beliefs, whereas sumo wrestlers and real-estate agents are probably typically conscious of the financial incentives at play. One way we can tell that financial incentives aren't the only ones that people care about is that financial incentives sometimes backfire. When Uri Gneezy and Aldo Rustichini teamed up with an Israeli preschool to deter parents from picking up their kids late, they found that parents picked up their kids even later when there was a fine. This is sometimes seen as paradoxical. After all, a fine increases the pecuniary costs of being late, and every economics model says demand (for picking up kids late) should decrease as the price goes up. But it's only paradoxical if one (wrongly) presumes that the price should only include pecuniary costs. As Gneezy and Rustichini point out, that's simply not true. There is a second cost to being late, social opprobrium, and this cost is reduced by the introduction of the fine. Similarly, expanding the definition of incentives to include other primary rewards can explain other seemingly paradoxical results, like the fact that we sometimes work harder when we are not paid for our work or that we believe the things we say more if we're not paid to say them.[2]

THE PROXIMATE-ULTIMATE DISTINCTION

If you google "Richard Feynman magnets," you'll find a video in which an interviewer attempts to ask the famous physicist why two magnets repel each other but gets, in response, an unsolicited lecture on the word *why*.[3] Here's Feynman:

When you ask why something happens, how does a person answer why something happens? For example, Aunt Minnie is in the hospital. Why? Because she went out, she slipped on the ice, and broke her hip. That satisfies people . . . but it wouldn't satisfy someone who came from another planet and knew nothing about things. . . . If you go, "Why did she slip on the ice?" Well, ice is slippery. Everybody knows that, no problem. But you ask, "Why is ice slippery?" That's kinda curious.

Feynman doesn't stop there. He describes why ice—a solid—is so curiously slippery, asks why some more, and then, finally, admits he's being rather obnoxious.

I'm not answering your question, but I'm telling you how difficult the why question is . . . "Why did she fall down when she slipped?" It has to do with gravity, involves all the planets and everything else. Never mind! It goes on and on.

OK, OK, we get it. Whenever we ask why, there are many answers we can give. The one we should give depends on what we're trying to learn or, as Feynman puts it later in his rant, on our level of analysis. If you ask why Aunt Minnie slipped, you might be interested in the mechanism by which ice becomes slippery (pressure), or you might be interested in Aunt Minnie's motivations—why the heck she went outside when it was icy. Likewise, if we ask why the waiter at the Jewel of India skips the bagel and goes for the balti curry, one answer is that he finds the former bland and the latter flavorful. Another is the one we gave in the last chapter—that in the waiter's homeland (or perhaps his parent's), spices helped to prevent foodborne illness.

Some levels of analysis have been given useful labels. Explanations that focus on the thoughts and feelings (bland! flavorful!) that go through someone's mind as he or she makes a decision are known as proximate. Explanations like the ones we focused on in the last chapter that get at the function of these thoughts and feelings are known as—you guessed it—functional. Sometimes such explanations are also called ultimate, not because you couldn't keep asking why but because it's understood you won't gain much by doing so (we already know why people don't like getting sick from their food—no need to focus on this for the purposes of understanding why they like spices).

The distinction between proximate and functional explanations originated with biologists. Why do peacocks grow very long tails? A proximate answer is: peahens find very long tails to be very, very attractive. But why did peahens evolve to find long tails so attractive? What's their function? (Uh, sorry, but the answer will have to wait until Chapter 6.)

Biologists typically know better than to stop at a proximate explanation like "peahens find long tails to be attractive." They naturally dig a bit deeper and try to uncover a functional explanation. Of course, the proximate may sometimes be interesting or a part of the story, but never the end. Never a satisfying answer.

Like biologists, we will find it helpful to dig deeper, past the proximate answers that, admittedly, come readily and naturally, and focus, instead, on functional explanations. When we asked why Native Americans treat corn with an alkaline solution prior to cooking it, we didn't stop at proximate answers—such as "because it tastes better that way" and "because that's how we've always done it"—but focused on the functional one: "because this increases the nutritional value of the corn." Likewise, when, in Chapter 6, we discuss why humans come to find conspicuous luxury goods like mosaics, gardens, and Rolex watches attractive, we won't stop with "because they're pretty" but try to uncover the function that leads us to find them pretty. When in Chapter 8, we tackle why people find it aversive to litter, we again won't stop with "because it's wrong" but ask what function such a belief might serve. You get the idea.

THE EMIC-ETIC DISTINCTION

One more distinction that might come in handy.

In 2017, the Star Trek franchise launched *Discovery*. The series was the seventh since the original series (or TOS, as it is known to

Trekkies) was launched in 1966. A lot changed in those fifty-one years. Back in 1966, nine out of ten Americans still had black-and-white TVs. Computer graphics had not yet been used in a major feature film or TV series: it would be over a decade before films like *Star Wars* and *Alien* employed CGI in earnest. Production budgets also increased dramatically. When it was launched, TOS's budget was so tight that the show's primary antagonists—an alien species known as Klingons—were made up using (no joke) shoe polish. Indeed, they became the show's primary antagonists precisely because their makeup was much cheaper and less time consuming to apply than another antagonistic alien species, the Romulans.

Of course, as the years progressed and budgets grew, the show's sets, battle scenes, and makeup all improved. The starships made use of more and fancier tech like holographic displays, made possible by CGI. The Klingons' appearance changed drastically: shoe polish gave way to ridged foreheads and snaggle teeth. The Romulans' appearance evolved as well. And so on.

Although the actual (and obvious) explanation for these changes was improved technology and bigger budgets, the show's producers and its fans often engaged in elaborate storytelling, known as "in-universe" explanations, to "trekoncile" the show's inconsistencies. Why did the Klingons grow ridges on their foreheads after TOS? Because they'd been infected by a virus. Why was the starship's bridge redesigned? Because the starship was modular, and its bridge could be replaced if damaged or in need of an upgrade. Why was the starship in *Discovery*, which was a prequel to TOS, filled with fancier tech than the starship in TOS? Because all that holographic technology had proved unreliable and had been removed. Here's a Trekkie explaining that last one in his own words:

During discussion of the repairs to the Enterprise in "An Obol for Charon" it is noted that one of the most problematic areas is the holographic comms system. Pike [the starship captain] tells his engineer to rip the entire thing out and go back to 2D display screens.

That may also explain why the bridge viewscreen on the Enterprise
is 2D and smaller; it could be a retrofit, a temporary repair even
that became permanent.[4]

Coming up with in-universe explanations may be fun, but
no matter how clever Trekkies' explanations for *Enterprise*'s tech-
nological regression, the true story remains that in the fifty years
between TOS and *Discovery*, CGI improved and budgets grew.
Nobody confuses the in-universe explanation for the true cause.

There's an analogous distinction that arises when we consider
the explanations people give for their own actions. For instance,
what if we ask why Jews wear kippahs—small round caps—on
their head? Objective, outside observers give one explanation: because unique headdress was imposed upon Jews by the Abbasid
Caliphate in 850, the pope in 1215, the Ottomans from 1577 on,
and so on, then rabbis decreed that those who were observant
should wear kippahs even if not required to do so. But that's not
the explanation that Jews themselves typically give. Here's what the
Orthodox sect Chabad has to say about kippahs on chabad.org:

> The tradition to wear a *kippah* is not derived from any biblical pas-
> sage. Rather, it is a custom which evolved as a sign of our recogni-
> tion that there is Someone "above" us who watches our every act.
>
> The Talmud relates that a woman was once told by astrologers
> that her son is destined to be a thief. To prevent this from happen-
> ing, she insisted that he always have his head covered, to remind
> him of G-d's presence and instill within him the fear of heaven.
> Once, while sitting under a palm tree, his headcovering fell off.
> He was suddenly overcome by an urge to eat a fruit from the tree,
> which did not belong to him. It was then that he realized the strong
> effect which the wearing of a *kippah* had on him[5]

Anthropologists have a term for such explanations. They
call them *emic* (pronounced ee-mic). Emic explanations are the

explanations given by those steeped in a culture, as opposed to those given by an objective observer, which anthropologists would term *etic*.

In this book, we're going to tackle questions for which many of us have emic explanations at the ready, just as Jews do for kippahs. Why don't Protestants follow the pope? Here are some emic explanations that Protestants themselves give to this very question:[6]

"There are no perfect humans, at least none I know, and this includes the pope. But he's a sterling fellow I hear."

"The simplest way to explain it is that Catholics believe they need a go-between between them and God while Protestants don't."

"The pope's authority is actually built on surprisingly uncertain ground."

Why do poets impose constraints upon themselves like iambic pentameter? An emic explanation is: "Iambic pentameter mimics the human heartbeat in its rhythm." If you've read Locke, you'll have emic explanations when we get to rights. If you're a wine snob, you'll have them when we get to aesthetics. Everyone has them when it comes to altruism and ethics.

In this book, we need to shift away from the emic perspective that we're used to. We might ask why Protestants, poets, Locke, wine snobs, and so on give the emic explanations that they give, but we won't treat those explanations as *the* explanation any more than we'd treat Trekkies'.

◆

SO THAT'S WHERE WE ARE HEADING. TO EXPLAIN ALL OF OUR SOCIAL puzzles, we will use game theory, but the game theory will often be hidden and will need to be interpreted through the lens of learning and evolutionary processes. The explanations we will be using game theory to uncover will all be of the ultimate (not proximate!)

and etic (not emic!) variety, focusing on primary (not secondary!) rewards.

In the next chapter, we finally start to use game theory, but a reminder: it's not yet a chapter on irrational human behavior. It's a chapter on neither humans or even behavior. It's about animal sex ratios—the ratio of males to females in a given species. What does that have to do with anythi . . . Oh, never mind, you'll see.

CHAPTER 4

SEX RATIOS: THE GOLD STANDARD OF GAME THEORY

CHARLES DARWIN PUBLISHED HIS SECOND BOOK, *THE DESCENT OF Man, and Selection in Relation to Sex,* in 1871, twelve years after the earth-shattering release of his *On the Origin of Species.* The first half of his new book—the part on the descent of man—argued that humans and apes indeed shared common ancestors, a fact that most of our readers probably take for granted but that Darwin's contemporaries still questioned. The second half of the book explored questions related to sex, like: Why are there different sexes in the first place? What are some common differences between the sexes? And why did these differences arise? All this, with the goal of arriving at evolutionary explanations, of course.

A few pages into this second part, Darwin took a long aside to explore what he termed "the proportional numbers of the two sexes in animals belonging to various classes." That is, the proportion of males to females in a species, which, today, is most often simply referred to as a species' sex ratio. "As no one, as far as I can discover, has paid attention to the relative numbers of the two sexes throughout the animal kingdom," he began, "I will here give such materials as I have been able to collect, although they are extremely imperfect."[1]

Darwin began by discussing the sex ratio in humans. "In England," he intoned, "during ten years (from 1857 to 1866) 707,120 children on an annual average have been born alive, in the proportion of 104.5 males to 100 females. But in 1857 the male births throughout England were as 105.2, and in 1865 as 104.0 to 100." He continued by discussing the sex ratio in England's various districts, in France, and among Christians and Jews, in each case concluding that the sex ratio was, very nearly, 1:1.

Next, Darwin considered racehorses. "Mr. Tegetmeier has been so kind as to tabulate for me from the 'Racing Calendar' the births of race-horses during a period of twenty-one years, viz. from 1846 to 1867; 1849 being omitted, as no returns were that year published. The total births have been 25,560, consisting of 12,763 males and 12,797 females, or in the proportion of 99·7 males to 100 females. As these numbers are tolerably large, and as they are drawn from all parts of England, during several years, we may with much confidence conclude that with the domestic horse, or at least with the race-horse, the two sexes are produced in almost equal numbers."

He also covered dogs, sheep, cattle, birds, and insects. (His discussion of moths was particularly charming. It included a table with counts collected from acquaintances all over England. "The Rev. J. Hellins of Exeter reared, during 1868, images of 73 species, which consisted of 153 males and 137 females. Mr. Albert Jones of Eltham reard, during 1868, images of 9 species, which consisted of 159 males and 126 females." And so on.) In nearly all cases, Darwin found the sex ratio to be very close to 1:1.

So, why are sex ratios roughly 1:1 at birth?

One answer, sometimes given by well-meaning, if misguided, high school biology teachers, is that a 1:1 ratio ensures that males and females each have mates when they reach sexual maturity. This answer is intuitively appealing in a society that emphasizes monogamy, but it falls apart under minimal scrutiny. For starters, in nearly every species, and at nearly every age, males are more likely than females to die (leaving a less balanced ratio when it comes time to mate). As Robert Trivers put it in a 1976 article, "males apparently have a tendency to suffer higher mortality rates than females. This is true for those dragonflies for which there are data, for the house fly, for some lizards, and for many mammals." It's also true for us humans: "During the first four or five years of life, also, more male children die than females; for example in England, during the first year, 126 boys die for every 100 girls—a proportion which in France is still more unfavourable," wrote Darwin.

Moreover, species vary in how monogamous they are. Some, like elephant seals, are highly polygynous. A single male usually mates with dozens of females.[2] Others—puffins, eagles, jackals, and even some fish—do pair bond, as biologists like to say, sometimes even for life.[3] The biology teacher's answer would suggest that elephant seals would have a lower sex ratio at birth—fewer males for every female. This is not the case: elephant seals, parrots, and penguins all have a sex ratio of roughly 1:1 at birth, just like Darwin's horses and moths and us (leaving many males unmated). Indeed, Darwin himself addressed this point: "with mankind," he wrote, "polygamy has been supposed to lead to the birth of a greater proportion of female infants; but Dr. J. Campbell carefully attended to this subject in the harems of Siam, and he concludes that the proportion of male to female births is the same as from monogamous unions. Hardly any animal has been rendered so highly polygamous as our English race-horses, and we shall immediately see that their male and female offspring are almost exactly equal in number."

Instead of our biology teachers, we can turn to the great statistician and biologist (and, sadly, eugenicist) Ronald Fisher. He is normally given credit for the answer, which has been called "the most celebrated argument in evolutionary biology."[4] The coolest part? This answer is effectively a game theoretic one, even though it was developed long before game theory was invented.

The intuition: only at a 1:1 sex ratio will no parent expect to have more grand-offspring by birthing one sex over the other. So they are fine sticking to fifty-fifty randomization. Let's drill down into this and see why and how it'd be formalized.

In Fisher's game, there is a population with two types of players: males and females. The size of the population won't end up mattering, but to make things concrete, you can imagine there are 100 players in all and that the population size doesn't change from generation to generation. The sex ratio of the population is the key outcome we're interested in. Remember that the sex ratio is defined as the ratio of males to females, so if there are 75 males and 25 females in Fisher's population, then the sex ratio is 3:1. If there are 25 males and 75 females, then the sex ratio is 1:3. What if there are 33 males and 67 females? Now the sex ratio is 1:2. You get the idea.

As in any game, we begin by specifying the actions players can take. In Fisher's game, players can choose the sex of their offspring. Of course, in reality, animals don't literally choose the sex of their offspring but bear with us for now.

Next, we must specify payoffs. In Fisher's game, it is easiest to think of payoffs as the number of grand-offspring that a player has; that is, the number of offspring the player's offspring have. And in case it isn't obvious, more grand-offspring is better.

Finally, let's make some assumptions—three, to be precise. Our first assumption: male and female offspring are equally costly to produce. Our second assumption: mating is exogamous (i.e., no inbreeding). Our third assumption: every male is equally likely to be chosen as a mate, as is every female. This allows us to focus on the number of offspring males and females are expected to have. Of course, our three assumptions don't always hold, but stating them up front helps us understand the bounds to which our analysis applies. Later, we'll check what happens when these assumptions break, which will prove quite useful.

We're ready to analyze Fisher's game. We'll start by checking whether, at each sex ratio, players are better off choosing to have more offspring of a particular sex. This requires figuring out how many grand-offspring each sex provides. Choose a sex ratio of, say, 1:3. With a population size of 100, this means there are 25 males and 75 females. Since the population isn't growing, together, as a group, the males have 100 offspring among them. That's 4 each, on average. What about females? They, too, have 100 offspring among them—the same 100 offspring. But since there are more females, this averages out to just 100/75, which equals 1.33 points per female. So, when the sex ratio is 1:3, players in Fisher's game are better off choosing to have male offspring because this results in 3 times as many grand-offspring, on average.

We can perform the same calculations for any sex ratio. For instance, when the sex ratio is 1:2, male offspring have 2 times as many grand-offspring. Not as good as before, but still better than females, so players would still choose to have male offspring. When the sex ratio is 1:1.5, male offspring have 1.5 times as many grand-offspring. Still better. And so on. So long as the sex ratio is less than 1:1, meaning males are relatively less common, it will be better to have male offspring.

What happens when females are less common? Say the ratio flips to 1.5:1. Now it's female offspring who yield 1.5 times as many grand-offspring, so it's better to have females. When the ratio further increases to 2:1, female offspring yield 2 times as many grand-offspring. At 3:1 it's 3 times as many. So long as the sex ratio is greater than 1:1, it's better to have female offspring; that is, it's always better to have offspring of the less common sex.

What happens at a sex ratio of 1:1? At 1:1, male offspring have the same number of grand-offspring as female offspring: 50 males have 100 offspring between them, or 2 on average, and 50 females have 100 offspring between them. Also 2 on average. Now there's no benefit to having male offspring over female offspring, or vice versa.

NASH EQUILIBRIUM

The most important concept in game theory is that of the Nash equilibrium. It was developed by the mathematician John Nash, whose biography is (mis)portrayed in the film *A Beautiful Mind*.

In a Nash equilibrium, each player does as well as he or she can, taking as given what other players are doing. If any player can benefit by deviating from his or her current strategy, while others don't change their strategy, then players aren't at a Nash equilibrium. If no player can benefit by (unilaterally!) deviating, then voilà, we're at a Nash equilibrium.

Notice that this definition involves players behaving optimally. But it involves a bit more than that. It requires that they behave optimally, given what others are doing, and vice versa, in a recursive, self-referential sort of way. That might at first feel a bit strange, but it's actually what gives the Nash equilibrium a lot of its bite—something you'll see as you use it over the course of the book.

Nash is precisely what we used when analyzing sex ratios. The only Nash equilibrium of the game we described involves 1:1 sex ratios. As we will often do throughout this book, we made this argument in steps: first by showing that other potential ratios are not Nash equilibria. Then by showing that 1:1 is.

Here's the argument again, in brief. First, suppose the ratio were below 1:1, so that females predominate. At these ratios, parents benefit by deviating: they want to have more male offspring than they're having. (Benefit here means having more progeny.) The same can be said about a ratio above 1:1: at those ratios, parents also benefit by deviating, though, this time, to having more female offspring. Only at 1:1 do parents not benefit by deviating one way or the other. Voilà! 1:1 is the *unique* Nash equilibrium.

ARRIVING AT A NASH EQUILIBRIUM
WITHOUT CONSCIOUS OPTIMIZATION

Until now, we spoke as if players in Fisher's game were consciously choosing the sex of their offspring. As you all know, that's not what happens. Instead, evolution does the choosing for them. Darwin, presaging Fisher, described the process thus:

> Let us now take the case of a species producing from the unknown causes just alluded to, an excess of one sex—we will say of males.
> . . . Could the sexes be equalised through natural selection? We

may feel sure, from all characters being variable, that certain pairs would produce a somewhat less excess of males over females than other pairs.

In other words, there is, naturally, some variation in the sex of parents' offspring, and for some parents, the deviation will be in favor of the less common sex (females, in Darwin's example). Those parents, Darwin says, are lucky since they will, on average, have more grand-offspring:

The former, supposing the actual number of the offspring to remain constant, would necessarily produce more females, and would therefore be more productive. On the doctrine of chances a greater number of the offspring of the more productive pairs would survive; and these would inherit a tendency to procreate fewer males and more females. Thus a tendency towards the equalisation of the sexes would be brought about.

As evolution optimizes, it shifts the population toward the Nash equilibrium of 1:1. How? The lucky parents don't just have more grand-offspring, each of those grand-offspring is more likely to be female and to have females. Over time, females become less uncommon, and the sex ratio shifts down toward 1:1. A similar dynamic plays out if we start out at a sex ratio of less than 1:1, when males are less common. Now, it's parents who have more males who are lucky. They have more grand-offspring and pass along the tendency to have more male offspring. Over time, the number of males grows, and the sex ratio shifts up toward 1:1.

Only when the sex ratio is 1:1 is it stable. Those who have more male offspring do no better and no worse than those who have more female offspring. Even if, by some chance, the sex ratio did change, it would only do so briefly, since parents who have a tendency to have offspring of the less common sex would quickly push it back toward 1:1. No wonder Darwin found a sex ratio of 1:1 everywhere he looked.

In every game we examine from this point on, we'll be looking for Nash equilibria of the game: for the place or places—as there can be more than one Nash equilibrium—where players get stuck because nobody can benefit by unilaterally deviating.

When looking for Nash equilibria, we're making an important assumption: that players' actions are chosen optimally; that is, if their current choice isn't optimal given what others are doing, they'll deviate, and if their current choice is optimal, they won't deviate. But, as we've tried to hammer home in prior chapters, that doesn't mean that players *consciously* optimize. The players in Fisher's game are racehorses, dogs, sheep, cattle, birds, and insects, none of whom consciously optimize very much at all, much less the sex ratio of their offspring. Evolution optimizes for them.

HOW TO TELL IF A GAME THEORY EXPLANATION IS THE RIGHT ONE

How can you tell if the game theory models we present are the right explanation? The best way to provide evidence for a game theory model is to break its assumptions and see what happens. You probably remember that Fisher's model had three key assumptions. Let's break them down one by one.

The first assumption we'll break is: male and female offspring are equally costly to produce. This means that, in equilibrium, they must yield an equal number of grand-offspring, which, as we saw, only happens at a sex ratio of 1:1. But what if females are twice as costly to produce? Well, then they'd need to yield twice as many grand-offspring for anyone to benefit enough from producing these uber-costly females. This will happen at a ratio of 2:1. What if they're three times as costly? They'll need to yield three times as many grand-offspring, which will happen at a ratio of 3:1. Half as costly? A ratio of 1:2. *Etc.*

The biologists Robert Trivers and Hope Hare were the first to realize that this assumption could be turned into a prediction—that it is a feature of the model, not a bug.[5] Rather than always predict-ing a sex ratio of 1:1, Fisher's model provides a more nuanced pre-diction, tying the relative cost of rearing the sexes to the sex ratio.

Trivers and Hare knew that in some species of ants, females are much bigger at birth and thus costlier to produce. Would these species prove to be exceptions to the 1:1 rule and have a higher sex ratio, as predicted? Trivers and Hare went out into the field to find out. They collected drones and queens as they flew out for their first mating flight from about two dozen species in all. They sexed and counted them to determine the sex ratio and were care-ful to exclude any individuals who don't play a role in producing grand-offspring. They also dried and weighed them, which gave them a rough measure of how much an individual of each sex costs to raise. Then they plotted their findings. Sure enough, they spot-ted a clear trend. When the ratio of the dry weights was 1:1, the sex ratio was about 1:1, but for species where females were heavier, the ratio rose to 2:1, 3:1, 5:1, and, in one case, all the way to 8:1.

As Trivers and Hare expected, the exception—to the equal cost assumption—proved Fisher's rule. Their fieldwork provided grade A evidence for Fisher's model because only Fisher's model predicts the relationship found—other potential explanations for 1:1 sex ra-tios do not. For instance, think back to the explanation sometimes given by well-meaning biology teachers. In it, males and females were paired up to mate when they reached sexual maturity, regard-less of how costly they were to produce. This explanation predicts a 1:1 sex ratio even for the insects that Trivers and Hare studied. Trivers and Hare's evidence thus casts further doubt on the naive biology teacher's explanation and on any other explanation that does not predict a relationship between the cost of offspring and the sex ratio.

The second assumption in Fisher's model is that mating is ex-ogamous: offspring aren't mating with each other. The ecologist Edward Herre was the one to turn this assumption into a predic-tion.[6] Herre looked around for the prime example of a species that mated incestiously, and he found it in the fig wasp. As their name suggests, fig wasps lay their eggs inside a fig. After the eggs hatch, the brood of offspring stay put in the fig, where they grow to maturity. When the time comes to mate, they have to mate with a wasp in their fig, and most of the time, that means mating incestuously with a sibling—about as far from mixing as it gets.

When offspring mate incestuously, how can parents maximize the number of grand-offspring? Have as many females as possible to carry the grand-offspring to term, and one or so males to fertilize those females; that is, they should basically minimize the male to female sex ratio. So, is this what happens?

Now it was Herre's turn to head to the field for specimens. He collected dozens and dozens of broods from thirteen different species of fig wasps. Sure enough, he found that so long as a single mother had laid eggs in a fig, the sex ratio was exceptionally low: in most broods, the sex ratio was between 1:20 and 1:8. The parents had indeed minimized the sex ratio.

Sometimes, two mothers laid eggs in a single fig. In these two-foundress broods, there's a bit of exogamy and a bit of incest, and so parents no longer benefit by minimizing the sex ratio of their offspring; the optimal sex ratio is somewhere between the minimum and 1:1. Sure enough, Herre found that the sex ratio in most of his two-foundress broods was between 1:7 and 1:5—still a far cry from 1:1 but quite a bit higher than in his single-foundress broods. When taken together with the evidence from single-foundress broods, this is slam-dunk evidence for Fisher's model since, again, other than Fisher's model, there aren't any other explanations for the 1:1 sex ratios that predict a relationship between the degree of mixing and the sex ratio.

The third assumption we made was that every male is equally likely to be chosen as a mate as is every female. This assumption is not realistic because some offspring are better at the mating game. In fact, parents might, to some extent, be able to anticipate how successful their offspring will be at this game, depending on how many resources they can provide their offspring or how good they themselves were at this game.

Trivers, together with Dan Willard, predicted that this, too, might affect the sex ratio any given parent would choose.[7] Their argument was that being successful relative to one's peers is helpful for females but not that helpful since biology tends to place something of an upper bound on how many offspring females can have. On the other hand, being successful relative to one's peers is extremely helpful for males: in most species successful males have many, many more offspring, while unsuccessful males have none. Trivers and Willard thought, What if successful parents tilted the sex of their offspring toward males, while unsuccessful parents tilted them toward females, to make the most of their respective advantage?

Dozens of studies have since tried to provide evidence for what is now known as the Trivers-Willard hypothesis. Some succeed, some don't, but on net, the evidence in favor of the hypothesis seems compelling. Elk, bison, mule deer, reindeer, roe deer, white-tailed deer, blue-footed boobies, lesser black-backed gulls, kakapos, pigeons, robins, finches, moths, and humans are just some of the animals where there is a known tendency for successful individuals to have more male offspring.[8]

But, at the end of the day, Trivers and Willard don't necessarily predict that the average sex ratio in the population will veer from 1:1, just that some parents might sometimes shift the ratio of their offspring from the population mean. If enough parents do this—say it's a boom time, and they have more males—then, eventually, Fisher's dynamics will kick in: other parents will shift their sex ratio toward females, until the sex ratio returns to 1:1, on average. Sure enough, in practice, a 1:1 sex ratio persists even in species where there is evidence for the Trivers-Willard hypothesis. In other words, the third assumption sometimes breaks, but even when it does, Fisher's result holds up, just not at the level of individual parents.

BEFORE WE CLOSE THIS CHAPTER, WE'D LIKE TO DISCUSS A FEW POINTS that will prove useful in our upcoming analyses.

COMPARATIVE STATICS

"The sex ratio will tilt more toward males if parents must invest more in females." "The sex ratio will tilt more toward females when there is more incest." "You're more likely to develop a passion for things that are more socially valued." "You're more likely to become passionate if you stand a chance at becoming the best." Statements like these are thought experiments in which we imagine ourselves approaching the model's control panel, finding the appropriate dial—the "parental investment dial," the "incest dial," and so on—and turning just that dial without touching the others. Such thought experiments are known as comparative statics—so called because they compare what happens when you change one thing while holding everything else static.

Comparative statics are the key way that economists make predictions, and they're the kind of predictions we'll make in this book. They're powerful, but as we hinted at in our discussion of primary rewards, they have their limitations. For starters, they're crude. A comparative static will not yield precise predictions like "if you give six-year-old Ledecky a pair of goggles, her chances of becoming a legend are 93.37 percent." They can only provide qualitative or directional predictions of the form "all else equal, more X is probably going to get you more Y, on average." That's what we expect from our models. Anything more would be taking the models too literally.

It's also not usually possible to turn just one knob without all the others also adjusting at the same time. That's why we run laboratory experiments and why there's a cottage industry of empirical economists and statisticians whose entire day job is finding clever natural experiments or instrumental variables. These are the statistical equivalent of turning just one knob.

THE OPTIMIZATION PROCESS
AFFECTS THE INTERPRETATION

While it's possible to justify using the Nash equilibrium in all sorts of ways, different optimization processes do require different interpretations. Here are two differences.

What is the currency? That depends on the force doing the optimizing. In this chapter, that force is natural selection, so the currency is fitness, which is proxied by the number of grand-offspring. When conscious choice is at play, which is not usually the case in this book, pleasure or some other conscious pursuit might be the appropriate currency. When market forces are doing the optimizing, profits will be the appropriate currency. In our book, the force doing the optimizing is usually learning, so primary rewards are the appropriate currency.

When should we expect a lag or spillover? Biological evolution is relatively slow. It takes generations to select between more or less optimal traits. If you put an organism in a new environment, it can take quite a long time for them to adapt. Learning processes adapt more quickly though not instantaneously. We expect learning to lead to some lags, but they won't stick around for as long as the whale's "hand" has. Conscious optimization, on the other hand, is nearly instantaneous. It happens as quickly as you can think, process, and calculate. So, lags won't be so relevant. And the same can be said about spillovers.

◆

FISHER'S MODEL SETS THE STANDARD WHICH WE ASPIRE TO MEET WITH all the game theory models in this book. It provides a simple, compelling explanation for a phenomenon that would otherwise be puzzling. It requires relatively few assumptions, which are usually reasonable. Then checking what happens when those assumptions break turned out to be a powerful source of predictions.

Such predictions often provide slam-dunk evidence—the kind that no alternative explanation predicts and the data unambiguously supports.

In the next chapters, we will finally use game theory to understand puzzling aspects of *human* behavior (as well as our corresponding tastes and beliefs). As we do this, we will do our best to try to make our case as compelling as Fisher's, by focusing on otherwise inexplicable puzzles, minimal robust assumptions, and key comparative statics.

CHAPTER 5

HAWK-DOVE AND RIGHTS

REMEMBER HOW, ONCE UPON A TIME, WE USED TO PAY FOR CABS WITH cash? At the end of the ride, the total would pop up on the meter, and we'd pull out our wallet, hand over a twenty-dollar bill, get our change, and be on our merry way.

Although this seems perfectly natural, as Kaushik Basu points out, it's actually pretty amazing that things go so smoothly.[1] Why did anyone ever pay? Why not just hop out of the cab and claim they've paid but without actually doing so? The taxi driver couldn't prove otherwise. And why doesn't the taxi driver pull a similar stunt and insist on a second payment? It's not like passengers can prove they've already paid. And it's not like the first twenty is any less valuable to them than the second. Yet, we're not as inclined to hand the second one over. Why not?

At a proximate level, we know the answers to these questions. The driver has a right to the first twenty but not the second. The first one's his, but the second is not. That one is ours. A suggestion to the contrary would likely lead to a dispute or even a fight. And that's just how it is. Who even asks these sorts of questions? (Uh, we do.)

Yes, clearly, we have strong intuitions over what's ours and what's not. But what does that even mean? This chapter is going to try to address that question and more with the help of the hawk-dove game.

THE HAWK-DOVE GAME REPRESENTS A CONTEST OVER A RESOURCE, like food, territory, a mate, a patent, oil fields, or a pig (yes, a pig). The game is, of course, a highly stylized representation of this contest. There are two players. Both choose from the same two strategies: hawk (be aggressive) and dove (acquiesce). For animals contesting food, this might literally mean a choice between fighting or walking away. For firms contesting a patent, it might mean litigating or settling. For countries contesting an oil field, it might mean choosing between going to war and ceding rights to the oil. The players choose their strategy simultaneously. This doesn't necessarily mean that they literally choose at the same moment—rather that they choose without first knowing the other's strategy.

The payoffs in the game—as in any game—depend not only on a player's own strategy but also on the other player's. If both players play hawk, they have an even chance of obtaining the resource, but a costly fight ensues. If they both play dove, they again have an even chance obtaining the resource but don't fight. If one plays hawk and the other plays dove, they also avoid a fight, and the one who plays hawk gets the resources. The other gets nothing. We typically assume that when a fight is sufficiently costly it is usually not worth fighting just for the opportunity of obtaining the resource; that is, the food, patent, or oil field in question is, of course, valuable, but not so valuable that it's worth a dangerous physical altercation, a lawsuit, or a war, respectively.

The same fundamental building blocks that made up Fisher's sex ratios game are present in the hawk-dove game—and every game. They are: *players* who play *strategies* that lead to *payoffs*. A player might be an animal, person, firm, or country. Examples of strategies, in addition to those already presented, are have sons, get angry, bid the true value of this band of spectrum to your firm, and blockade Cuba. Payoffs represent success on some dimension and can, for example, be denominated in offspring, prestige, dollars, or power.

This terrine of players, strategies, and payoffs can be characterized as a game because players' payoffs depend not only on their own strategy but also on what the other players do. That's the defining feature of a game. That's what makes games hard to study using standard optimization approaches and why a special branch of mathematics arose just for games.

We can see the key problem that games pose when we try to analyze what's optimal in the hawk-dove game: each player's optimal strategy depends on what the other chooses and vice versa. How do you solve such a circular problem? Thankfully, John Nash solved it by directing us to first specify a strategy for everyone and then check if any player can do any better, *assuming others are acting according to the specified strategy*. If no one can benefit by deviating to another strategy, we have a Nash equilibrium.

What are the Nash equilibria of the hawk-dove game (yes, there can be more than one)? Well, we have to check all possible pairs of strategies: both play hawk, both play dove, one plays dove and the other plays hawk, and vice versa. For each, we ask: Could either player benefit from deviating?

Let's start with if they both play hawk. Could a player benefit by deviating? Yes. Either would do better if they switch to dove. In that case, the player who switches gets none of the resource, but at least she doesn't get into a costly fight. So both playing hawk ain't Nash.

What about both playing dove? No dice. Any player who deviates to playing hawk will get the resource for himself and still avoid a fight.

What about if one plays hawk and one plays dove? Bingo. Neither can benefit by deviating. The one who plays hawk gets the resource without a fight. If she deviates to dove, she would have to split the resource. The one who is playing dove is stuck without the resource but would get stuck in a costly fight if he deviates to hawk. Not worth it. He ends up with the raw end of this deal, but he still can't benefit by deviating. He also can't count on the other player simultaneously deviating. That might be nice, but that's not how Nash works. So we've found the Nash equilibria of the game: (hawk, dove) and (dove, hawk). They're not fair, but they're Nash nonetheless.

. Hidden in this seemingly ho-hum revelation is the key insight of this chapter: so long as one player is expected to play hawk and the other dove, neither has a benefit from deviating. It doesn't matter which player is expected to do what. It doesn't matter if there's a good reason for this expectation. All that matters is that it's what everyone expects. Expectations, no matter how unjust or ill-justified, are self-fulfilling.

THE FIERCE SPECKLED WOODS OF OXFORDSHIRE FOREST

The hawk-dove game was developed by John Maynard Smith and George Price to explain animal territoriality.[2] The puzzle was that when animals fought over a resource—be they ducks, dogs, or deer—very often, the fight was symbolic: the animal that arrived first defended the resource ferociously, while the second animal, after testing the waters, retreated. A head butt or two, a few seconds of vigorous flapping of the wings, and the conflict was over, the incumbent inevitably victorious and the newcomer beating a hasty retreat. Why wasn't the fight more serious? Why did the incumbent always win? The hawk-dove game answered both questions. So long as the cost of the fight was high (namely, higher than the expected value of splitting the resource), then, in equilibrium, one animal would be expected to be aggressive and the other to acquiesce.

But which one? The order in which animals arrive influences both animals' expectations over who will play hawk. Both expect the animal who arrived first to be aggressive. As we just saw, once in place, this expectation is self-fulfilling. It is in the best interest of the animal who arrived first to be aggressive because she expects the other to acquiesce, and it is similarly in the interest of the animal who arrived second to acquiesce because he expects the other to be aggressive.

In 1976, soon after Smith and Price developed the hawk-dove game, Nick Davies released a study providing evidence for Maynard Smith's hypothesis from an unlikely source: the speckled wood butterflies that were abundant in the woods surrounding his hometown of Oxford, England.[3] The male speckled woods, as he affectionately called them, spend the bulk of their day in a single sun spot, staying warm and waiting for females to approach. As the sun moves across the sky, and the sun spot across the forest floor, the butterflies follow it. Here's how Davies describes it:

> Speckled woods spent the night high up in the tree tops, 5 to 15 m off the ground. The first signs of activity were seen early in the morning as the warmth of the sun reached the foliage of the woodland canopy. The butterflies opened their wings and orientated towards the sun to warm up. Between 07.00 and 08.00 hours B.S.T. a few could be seen flying about the tree tops and then gradually, over the next hour or two, the males descended to ground level where the sun's rays began to cast pools of sunlight onto the woodland floor. From this time on, throughout the day until early evening, males could be seen fluttering in these sunny pools. It was rare to see any in the shady part of the wood and it seems likely that the sunny spots were preferred because it was here that there was sufficient warmth to enable the butterflies to remain active . . .
>
> Individual males often spent the entire day in one sunspot. As the sun moved across the sky and the spot moved, they followed it, always keeping within the boundary. During the day males would move up to 50 m in this way as they followed their sunspots' travel across the woodland floor.

The males use the sun spots to keep warm while they wait on a prominent perch for females to come by. When one does appear, the male courts her, and if she fancies him, they head together to the canopy to mate. Of course, females aren't the only ones who do come by. Sometimes a bug from another species wanders over. When this happens, the male wood speckled inspects and then ignores it. And sometimes, another male speckled wood comes down from the canopy in hopes of securing the sun spot for himself. That's when all hell breaks loose (in butterfly terms, that is). Here's Davies again:

> Whenever another male speckled wood flew past, a spiral flight took place. Both individuals fluttered close to each other in mid-air, appearing to bump into each other, and spiralled vertically upwards towards the tree canopy. Then after a few seconds one of them turned and came down again to perch in the sunspot, while the other flew up into the tree canopy.

At first, Davies thought the spiral flight was "a means of territorial defence," but he soon noticed it wasn't a real fight:

> Out of 210 spiral flights between males where I had the owner marked, it was the owner which won the contest on every occasion. This applied even when the owner was a very tatty male whose wings were torn and the intruder was a perfect male in mint condition. Thus the spiral flights, only a few seconds in duration, cannot be regarded as contests at all. Rather they are a short conventional display where, to put it anthropomorphically, the owner says 'I was first here' and the intruder says 'Sorry, I didn't know there was anyone occupying the sunspot, I'll retreat back to the canopy'.

Davies began to suspect that Maynard Smith's hawk-dove model might apply. The male speckled woods could be thought of as players in the game, contesting the sun spot. The model's two key assumptions seemed to be met. Sun spots qualify as a valuable resource since they're where females go to find mates. But each particular sun spot is not *that* valuable; there are plenty of other sun spots in the forest. Moreover, fights are costly. As Davies put it, "a long contest is probably costly, both in terms of wasted time and energy and also perhaps in the risk of physical injury, such as wing damage during the spiral flights." Consequently, the speckled woods seem to have evolved intuitions that led them to play hawk-dove, where the owner of the sun spot was expected to play hawk.

Davies verified his suspicion in a number of ways. He wondered, for instance, why intruders ever entered the occupied sun spots in the first place if they expected to play dove upon entering. The answer was that they probably didn't do so intentionally: an intruder entered an occupied sun spot every eight minutes or so, but they entered unoccupied ones about twice as often.

He also wondered if a fight would break out if both males thought they had arrived at the sun spot first. To check this, he needed to sneak a male into an occupied sun spot without either male noticing the other. This proved rather tough, but ultimately, Davies was able to induce some genuine fights, which lasted an average of 40 seconds, instead of the usual 3.7 seconds of the spiral flight. Here's Davies again:

> I attempted to release a second male into an occupied territory without the resident noticing. This was rather difficult to do and on most occasions I failed because the owner spotted the second male before he landed in the sunspot, flew over to him and chased him off with a brief spiral. However, on five occasions I managed to put an intruder onto a territory unnoticed so that two males were perching together in the same sunspot. The first individual to fly was quickly spotted by the other and a spiral flight occurred . . . [which] was over 10 times as long in duration as the normal spirals when the roles of resident and intruder were clear to the contestants. Thus when both males think they are the resident, an escalated contest occurs.

Finally, Davies wondered whether an alternative explanation might be at play: Perhaps the sun spot's owner had an advantage in defending it? To test this, Davies tried netting the sun spots' original owners after they'd been there for some time and then waited for another butterfly to arrive and occupy the sun spot. After it arrived, he waited just ten seconds—hardly enough time for the new owner to develop a home court advantage—before releasing the original owner into the sun spot. Sure enough, the new owner won every time:

> On 10 occasions, each in a different sunspot with a different male, I removed the owner and held him in a net. Within a few minutes a replacement male arrived and took over the territory. I allowed the replacement to land in the sunspot and perch there for 10s. Then I released the original owner back onto the sunspot. The replacement immediately flew over to the owner and the two males engaged in a spiral flight up towards the tree canopy. After a few seconds they parted. One flew up to the canopy and one returned to the sunspot. On all 10 occasions it was the original owner who retreated . . .
>
> Although the original owners always lost their territories to the replacements, I thought that at least they would not give in without a good fight. So I compared the duration of the spiral flights between the replacement and the released owner with that of other contests . . . there were no significant differences. In other words, not only does the original owner lose the contest, he retreats without much argument!

These observations strongly suggested that the rule for deciding the territorial disputes was 'resident wins, intruder retreats'.

PROPERTY RIGHTS IN HUMANS

Like ducks, dogs, deer, and wood speckled butterflies, humans use "I was here first" to determine ownership. Get to a theater seat first and put your jacket down on it? It's yours. Doesn't matter if the guy who comes next is a lot bigger. Got the last red velvet at Georgetown Cupcake? It's yours. Doesn't matter that the guy behind you drove an hour and stood in line for another hour to get it. Who arrived first is enough to determine rights.

Peter DeScioli and Bart Wilson have a nice laboratory experiment illustrating this very point. The experiment took the form of a simple game set in a universe with scarce resources. In the game, subjects, represented by avatars, ran around collecting berries and would die if they didn't consume enough of them. Berries grew on bushes, which only one avatar could pick from at a time. Avatars could try to dislodge one another, if they both wanted to eat from the same bush at the same time, by throwing punches. Avatars came in different sizes. Some subjects got lucky and theirs was big and burly; they could throw meaner punches and would be more likely to win a fight. Others had puny avatars, which could have been modeled after the authors of this book.

When DeScioli and Wilson dug into the data to see what determined who ended up with the berry bush, they found that, typically, it wasn't size or even subjects' health levels that mattered. Instead, it was whoever arrived first. The subject who arrived at the bush first was typically the one there after the fight (er, "fight"). Their subjects were behaving just like ducks, deer, and speckled wood butterflies.

The intuitions DeScioli and Wilson document develop early. Consider the following study done on toddlers, who were brought into the lab, shown cartoons of children playing with a ball, a doll, or another toy, and asked, "Whose ball is this?" The toddlers consistently pointed to the cartoon figure who had the ball first.

Once again, determining who arrived first can have this effect because it influences people's shared expectations over who will aggressively defend the berry bush, the theater seat, the ball, and so on. If you arrive first then, usually, people won't demand that you give it up. And that's because they don't expect you to give it up easily. Meanwhile, you wouldn't give it up easily because you don't expect anyone to push very hard for it. Everyone has a shared sense of who has the right to the berry bush and theater seat and every incentive to behave accordingly.

◆

WHO ARRIVED FIRST IS AN EXAMPLE OF WHAT GAME THEORISTS CALL an uncorrelated asymmetry. It is uncorrelated because it is not directly related to the payoffs of the hawk-dove game. Which player arrived first tells you nothing about who is hungrier and needs the resource more or who is stronger and thus more likely to win a fight or who is healthier and thus more likely to survive if injuries are sustained. It is an asymmetry because it differentiates the players: one arrived first and the other didn't.

Who arrived first is just one of many uncorrelated asymmetries that we rely on. Another particularly important one is current possession. Whether a ball is in your hand or mine is uncorrelated with its value, but it is an asymmetry that we can rely on to establish shared expectations of who will behave aggressively over the ball.

Infants, for instance, seem to primarily infer ownership from possession. It's only later (at around age three) that ownership and possession become divorced, and they reliably infer that the first person who they saw possess an item (the one who arrived first) is the owner of the item, even if the item is not currently in their possession.[4]

Possession is also, famously, nine-tenths of the law, by which people mean that ownership often defaults to those who possess an item. This principle even played a role in triggering the famous feud between the Hatfields and the McCoys (which we'll return to in Chapter 14). At the very beginning of the feud, the families went to court over a sow and her piglets (see, we told you there'd be a pig). Neither side could decidedly prove their case, so the judge sided with the Hatfields, citing their possession of the pig as his justification for the ruling. This principle isn't unique to

Anglican law code. In Roman law, so long as it couldn't be shown that an object had been stolen or otherwise obtained via illegitimate means, then he who had *possessio* over the object also held *dominion* (ownership).

Another commonly used asymmetry is who built it. Indeed, even birds use this one. In many species, if a bird leaves its nest to hunt or forage and returns to find that a Goldilocks has taken its place, the Goldilocks will quickly cede the nest. In this case, who built it established the birds' shared expectations.

Humans, too, seem to heavily rely on who built it. It is common for rights to a parcel of land to be granted to the individual who builds on the land and improves it by adding fences, irrigation, and so on. This is known as the homesteading principle. Here's Locke's (emic) justification for it:

> Though the earth and all inferior creatures be common to all men, yet every man has a *property* in his own *person*. This nobody has any right to but himself. The *labour* of his body and the *work* of his hands, we may say, are properly his. Whatsoever, then, he removes out of the state that Nature hath provided and left it in, he hath mixed his *labour* with it, and joined to it something that is his own, and thereby makes it his *property*.

Ultimately, this uncorrelated asymmetry found its way into a variety of laws, including the Homestead Act of 1862, which was signed into law by President Lincoln and responsible for the settlement of much of the American West. It was also used to justify the accompanying displacement of the West's indigenous residents, whose own *labor* to improve those lands (managed fires, large and permanent buffalo corrals, and so on) was, conveniently, not recognized.

What are some other common uncorrelated asymmetries we use to establish shared expectations? Again, we can glean some from studies of children. For instance, by age four, children understand transfers of ownership like buying and giving and differentiate these from stealing; that is, at this point, they are able to use "was it paid for" and "was it given" to determine rights. (The uncorrelated asymmetry "was it paid for" prevents fistfights from breaking out in Myerson's taxi cab example from earlier: if you already paid for the cab ride, the cab driver doesn't demand you pay again because he doesn't expect you to give in to such a demand and you wouldn't give in to such a demand since you don't expect a cab driver to push you on it.)

Alternatively, we can look at the law—the remaining one-tenth of it. Peter DeScioli and Rachel Karpoff showed subjects brief descriptions of classic property rights cases, such as these:

- *Armory v. Delamirie:* Armory found a ring in the street and brought it to a goldsmith to be appraised. The goldsmith removed the jewel from the ring and refused to return it.
- *Bridges v. Hawkesworth:* Bridges found an envelope full of cash on the floor of Hawkesworth's shop.
- *McAvoy v. Medina:* McAvoy found a wallet with cash on a countertop in Medina's barbershop.

Then, DeScioli and Karpoff asked their subjects who, in their opinion, owns the item and also to describe why they made this choice. They found that subjects repeatedly and consistently relied on determinants like who lost the item, who found it, whose land was it on, in addition to who made it and who possesses it.

AS THE DESCIOLI AND KARPOFF STUDY ILLUSTRATES, OUR INTUITIONS about property rights often correspond. For instance, in *Armory v. Delamirie*, 100 percent of subjects chose Armory (the finder) as the rightful owner. That's what you would expect and what the hawk-dove story requires—*shared* expectations.

But this study also illustrates the potential for conflict, namely when there are multiple uncorrelated asymmetries floating around. In *Bridges v. Hawkesworth*, Bridges (the finder) got 84 percent of the vote. In *McAvoy v. Medina* subjects were nearly evenly split. Like the two butterflies who each thought they had a right to the sun spot, these are the cases that can lead to real fights. In fact, they did. After all, the study is based on real court battles.

The battles can also take place on actual battlefields. When Argentina and Great Britain went to war over the Falkland Islands in 1982, both countries had long-standing claims to the territory. Argentina's went all the way back to 1816 when Spain granted it dominion over colonies it abandoned in the Pacific. But Argentina hadn't immediately acted on that option, and in the meantime, a small British colony developed there and prospered. Great Britain's claim, then, was based on longtime possession. The Israeli-Palestinian conflict similarly has conflicting uncorrelated asymmetries. Israel's claim is, at least partially, based on first possession: the Jewish people resided in those lands thousands of years ago. Palestinians' claim is more like that of the British over the Falkland Islands: two thousand years ago was a long time ago and what really matters is who was living there over the past few hundred years. But some Jews remained! And the Palestinians left! But they only left at gunpoint! And so on and so on. Lots of different uncorrelated asymmetries. And no end in sight.

WE'RE NOW GOING TO MOVE ON FROM OUR SENSE OF PROPERTY RIGHTS and use the hawk-dove game to shed light on some additional topics.

GENERALIZING BEYOND PROPERTY RIGHTS

So far, we have focused on people's intuitions regarding property rights. However, people proclaim rights that are far more abstract, like the right to free speech, to bear arms, to have health care, and so on. There is obviously a lot going on when it comes to these rights—lots of interesting questions to explore. Where do these rights come from? Why these particular rights? What about the expanding circle of rights—the idea that a set of rights initially granted to one group, such as the right to vote, tends to gradually be extended to other groups? Why does that happen? The hawk-dove model does not hold the key to answering these questions.[5]

There is, however, one aspect of these rights that the hawk-dove model can help us with. The main feature of the hawk-dove model is the importance of shared expectations, which, once established, are self-fulfilling. This feature appears to be present not just for rights over a ball, a dollar bill, a ring, and so on, but also more abstract rights like free speech or the right to bear arms. When people believe in these rights, they will fight for them because they expect others (or the government) to back down, and those others (or governments) will back down because they expect people to fight for them.

Here are some implications of this claim.

For shared expectations to support rights, rights need not be grounded in anything logical. "We hold these truths to be self-evident" is as good a justification as any. Of course, people do provide other justifications for the particular rights upheld in their culture, calling upon philosophers like Locke or religious writings when they do so. But these are emic justifications, and they are not needed to understand why rights are self-sustaining.

Shared expectations can also differ across cultures and over time. And so do rights. In most countries, people do not have a right to free speech or to bear arms. Shared expectations were set differently in those countries than they were in the United States. The hawk-dove model doesn't tell you why they were set differently, but it does tell you that whatever expectations were set in each location can be self-sustaining. Another example: rights within a marriage vary greatly across cultures and over time, as hinted to by the traditional Hebrew words for *my husband* (*ba'ali*) and *my wife* (*ishti*). These translate to "my owner" and "my woman," respectively, and hark back to a time when women were treated as property. Today, some secular Hebrew speakers might still have some sense of ownership in a romantic relationship ("Hey, what are you doing talking to *my* girlfriend?"), but most would find even this to be fuddy-duddy, and if they didn't, they'd likely find themselves without a girlfriend. Shared expectations have shifted and are continuing to shift.

Documents like the Declaration of Independence and the Constitution can help to make rights self-sustaining. Writing rights down in such documents, having them signed by respected leaders, distributing them widely, teaching them to children in school, and so on—all this helps to establish shared expectations. Once established, it can be very difficult to change these expectations. Gun violence in the United States is very difficult to address because the right to bear arms is enshrined in the Constitution's Second Amendment. There are a lot of people willing to fight for that right, with the expectation that others will back down. And they have a lot of guns.

Historical precedent can similarly make rights self-sustaining. The classic legal cases in DeScioli and Karpoff's study are classics because, once they settled a property dispute in a particular way, it became expected that similar disputes be settled in the same way going forward. The same can be true for more abstract rights, like the right to discriminate. Cases about whether a baker should be allowed to refuse to sell her cakes to a same-sex couple are not in the news because anyone cares about cakes but because of the precedent these cases set for discrimination in more important contexts—housing, labor, and so on. Precedent also played a meaningful role in the United States' gun control policies. In *District of Columbia v. Heller*, the Supreme Court interpreted the Second Amendment as applying to all individuals, not just those associated with the military or a militia. The ruling not only struck down

DC's handgun ban, but it also set a precedent for future efforts to regulate guns. And, as the hawk-dove model helps to clarify, once set, such a precedent is self-sustaining.

APOLOGIES

November 25, 2011, marked the beginnings of a particularly troubled period in the perennially troubled alliance between the United States and Pakistan. In an operation against the Taliban, a NATO force entered Pakistani territory without authorization and accidentally killed twenty-four Pakistani soldiers. A furious Pakistan responded by closing the supply routes to the United States' military bases. The resulting logistical morass cost the United States some $66,000,000 a month. The US initially refused to apologize because the Defense Department felt Pakistan shared responsibility for the error. A string of mishaps and affronts further held up the apology. Finally, in July of 2012, a full seven months after the crisis began, US Secretary of State Hillary Clinton called Pakistan's foreign minister to say, "We are sorry for the losses suffered by the Pakistani military." Within hours, the supply lines reopened.[6]

We've all found ourselves in the same shoes as Pakistan and the State Department, insisting on an apology from a friend, colleague, or family member, or agonizing over whether to provide one and how to phrase it. Why? Why do mere words like "I apologize" and "I'm sorry" carry so much weight in the first place?

The hawk-dove game suggests that we put so much weight on these mere words because these mere words are able to establish shared expectations. An apology from the United States to Pakistan makes it clear to both parties that the US cannot continue to use Pakistani skies at whim, without permission or penalty. An apology from a friend, colleague, or family member similarly makes it clear to both parties that the offense cannot be repeated without consequence.

This perspective also provides some nuanced guidance on when to apologize. The web is awash with popular science articles that unequivocally advocate for apologizing, with titles like "The Power of Apology." Then again, some people think you should "never apologize, mister. It's a sign of weakness!" as Captain Nathan Brittles (played by John Wayne) puts it, in the classic Western *She Wore a Yellow Ribbon*. Whose advice should you follow? Well, apologies certainly have their benefits: an apology makes it more likely that others will forgive you and the relationship will be repaired. But we can now more clearly see their cost: others will expect you to change your behavior, and if you don't, they are liable to get especially upset. Is the benefit worth this cost? Sometimes. That depends on how valuable the relationship is and how much you lose by changing your behavior. The game theory adds nuance that the pop science articles (and, um, classic Westerns) lack.

SHOULD YOU BE MORE ASSERTIVE?

Speaking of advice, here's some more advice we often encounter: Be more assertive. Make your voice heard. "Lean in!" as Sheryl Sandberg put it. If you don't find this to be so natural, well, then there are exercises you can do to help. In her TED talk "Your Body Language May Shape Who You Are," Amy Cuddy promotes spending just two minutes in "higher power poses" (for instance, "manspreading" or leaning back in your chair with your feet on the table) before important interviews and meetings, as a way of pumping yourself up so that you can be assertive, make your voice heard, and lean in.[7]

Is this good advice? If you're thinking functionally, you might already be asking yourself, "If this is such a good idea, why don't people then already do it all the time? Maybe there is a cost?" As with apologies, the hawk-dove model can help us see what that cost might be. Being assertive can mean playing hawk when others expect you to play dove and getting into fights.

Of course, sometimes people are expected to play hawk and don't. Maybe they have a natural inclination toward self-doubt due to a history of discrimination. Maybe they were bullied as a kid. Or maybe there's just something about their personality that leads them to be cautious or quiet. Maybe those inclinations are, quite understandably, spilling over from contexts where they worked to the boardroom, where they're backfiring. In such cases, being more assertive might indeed be good advice. These folks should take a deep breath, manspread a bit, maybe make some funny faces, and . . . lean in!

There are also cases when getting into fights might be worthwhile. Sometimes, people are willing to put up a fight to change people's expectations—to get into "good trouble," as Representative John Lewis would have put it. Such people might benefit by psyching themselves up with Dr. Cuddy's tricks before walking into the boardroom, just as an athlete does before walking onto the field.

In short, it's not black or white. Sometimes it's good to be more assertive, and sometimes not. The hawk-dove model can help you figure out when.

STOCKHOLM SYNDROME AND INTERNALIZED RACISM AND SEXISM

The term *Stockholm syndrome* was coined after the hostages in a failed Stockholm bank robbery refused to testify on behalf of the police and, instead, raised funds to aid in the defense of their captors.[8] Though seemingly irrational, the Stockholm hostages' response is quite common.

Two explanations for Stockholm syndrome are commonly given. The first is that kidnappers brainwash their hostages. The second is that hostages develop empathy for kidnappers after spending a lot of time with them, under especially tense conditions. Both of these explanations are, at least to an extent, clearly true, but why does the brainwashing work? And if empathy is automatically triggered by spending lots of time with people under tense conditions, why aren't kidnappers equally as likely to develop deep empathy for their hostages (this is sometimes called Lima syndrome, but it is much less common than Stockholm syndrome).

Before we use the hawk-dove model to address the shortcomings in these commonly given explanations, let us introduce another seemingly disparate phenomenon.

In the doll test, conducted in the 1940s, a psychologist placed two dolls—one with light skin, the other with dark—before a young child and asked a series of questions: Which doll is nice? Smart? Which is good? The sessions took a heartbreaking turn when Black children pointed to the white doll as the nice, smart, and good one. In the Supreme Court's 1954 decision in *Brown v. Board of Education* that ruled against segregation in schools, Chief Justice Earl Warren cited these results when he wrote that Black children suffered "a feeling of inferiority as to their status in the community that may affect their hearts and minds in a way unlikely to ever be undone."[9]

The doll test documented that even children can internalize the racism they observe and experience and begin to believe that they themselves are inferior. Sexism can likewise be shown to be internalized. Victims of domestic violence often argue their partner was justified in using violence against them, and in cross-cultural surveys, it is common for a third or more of women to respond that violence against women is "an appropriate way to deal with women's misconduct."[10]

Although the internalized racism and sexism may, like Stockholm syndrome, initially seem surprising (and disheartening), these responses to violence, abuse, and discrimination may be the best that those who have been kidnapped, discriminated against, and abused can do. In settings where playing hawk is always expected to get you beat, or worse, thinking of yourself as inferior or as deserving of your fate and of abusers or kidnappers as deserving of empathy or even worship can be self-preserving. As one kidnapping victim, Natascha Kampusch, argues about Stockholm syndrome: "[It] is not a syndrome. It is a survival strategy."[11]

We should, of course, acknowledge that internalized racism, sexism, and Stockholm syndrome aren't just enforced in a self-fulfilling way. In the hawk-dove model, players are equally likely to win a contest, whereas when it comes to racism, sexism, and hostage situations, there are real power asymmetries! But the model still teaches us that our sense of rights and worth is responsive to whether it's optimal to behave aggressively. Unfortunately, when aggressive demands are sanctioned, it's best to avoid making those demands. Internalizing a lower sense of worth can help with that.

	Hawk	Dove
Hawk	$\dfrac{V}{2} - C$	V
Dove	0	$\dfrac{V}{2}$

SETUP:

- In the standard hawk-dove game, there are two players. Each chooses from two actions: hawk (H) and dove (D).
- The payoffs to player 1 are presented in the matrix above, in which, by convention, player 1 chooses the row, and player 2 chooses the column.
- $v > 0$ represents the value of the resource being contested, $c > 0$ represents the cost of the fight that occurs when both players play hawk.
- The payoffs to player 2 are determined the same way, and are thus omitted from the payoff matrix.

STRATEGY PROFILE OF INTEREST:

- (H,D) and (D,H): player 1 plays hawk and player 2 plays dove, and vice versa; represented in bold.

EQUILIBRIUM CONDITION:

- So long as fighting is sufficiently costly relative to the value of the resource, $v / 2 < c$, (H,D) and (D,H) are the only Nash equilibria of the game. (Technically, these are the only pure Nash equilibria; in this book, we do not focus on mixed Nash equilibria—those in which players randomly select from amongst different actions.)

INTERPRETATION:

- Who gets the resource can depend on arbitrary events, like who arrived first, provided this arbitrary event affects everyone's expectation of who will play hawk. That is, in the hawk-dove game, expectations can be self-fulfilling.
- Which events players condition on isn't specified by the model and can depend on many factors such as context, culture, precedent, and efficiency considerations.

CHAPTER 6

COSTLY SIGNALING AND AESTHETICS

THE ANCIENT ROMANS LOVED THEIR MOSAICS AND CERAMICS. THE Persians, their gardens. In the fifteenth century, the burgeoning middle class of Europe was all aflutter over a light blue called pastel, laboriously made by fermenting a flower (*Isatis tinctoria*) grown in the southwest of France; the region grew so wealthy from trade in pastel that it became known as the Land of Plenty. In the sixteenth century, the gentry of Western Europe obsessed over fine tapestries. Meanwhile, the Chinese, Koreans, and, eventually, Japanese developed equally fierce obsessions over pottery, silks, and painted screens. Today, we faun over authentic Danish mid-century-modern furniture, wax poetic over the laboriously made product of another flower, saffron, sing songs (literally) about Lamborghinis and Moet fizzy, dish out hundreds for a pair of Yeezy sneakers or a Supreme T-shirt, and dish out many times more for a Rolex. Then there's jewelry, which was all the rage long before the Romans and still abundantly adorns many of the fingers and wrists that turn these pages. While the details vary across time and culture, one thing is constant: we humans have a penchant for really, truly digging fancy stuff.

Why? Mosaics are no more effective a ground covering than regular tile. Gardens are pretty, sure, but couldn't the Persians have amused themselves with a pastime that didn't involve intensive use of precious water? Pastel, too, is pretty, but nobody seems to be too worked up about blue dyes—pastel or otherwise—now that we've found many other ways to make them. As for tapestries, most people would find it pretty fuddy-duddy if you hung your entire living room with them, but that's exactly what Mary Queen of Scots did—and was enthusiastically complimented for. Lamborghinis go just as fast as Hondas in bumper-to-bumper traffic. Yeezies are no more comfortable than Clarks. A Supreme T-shirt is typically printed on the same material as the ten-dollar T-shirts hanging from the racks at Walmart. A Rolex tells time slightly *worse* than a forty-five-dollar Casio G-Shock and is substantially less tough. The gold and jewels tucked into King Tut's mummy wraps, on display in the Tower of London, or neatly laid in readers' jewel boxes all do nothing but gleam and sparkle. The stuff we dig may be fancy, but it's kinda useless, and we might genuinely dig it, but that kind of begs the question why.

What drives our timeless taste for luxury?

Of course, whether people are aware of it or not, mosaics, gardens, Lamborghinis, and Rolexes serve to show off wealth. As we'll see, the game theory will confirm this intuition. It will also help us predict when luxuries will go in and out of style. But there's something perverse about our penchant for luxury: Why does anyone admire these luxurious signals of wealth if they just serve to blow the very wealth they show off?

Darwin and other biologists of his time noticed an analogous puzzle.[1] In many species of birds, males grow long, ostentatious tails. The longer and more ostentatious the tail, the more attractive the male. Shorten a male's tail, and the unfortunate male will end up with far fewer nests that mating season. But he'll also be measurably better at hawking (catching bugs) and measurably nimbler as he flits about in the sky, rueing his lonely state of affairs. Why would females go gaga for a tail that made their mate—and would make their own offspring—less good at hawking and evading predators?

The costly signaling game was developed (simultaneously and separately) by Michael Spence and Amnon Zahavi to answer such questions. In this chapter, we'll present the game and use it to understand these puzzles.[2] We'll then apply it to a host of additional puzzles, like: Why do men in some cultures grow very long pinky nails? Why do artists impose artificial constraints like iambic pentameter on their art? And why do some religious groups (like some ultra-Orthodox Jewish sects) require members to engage in seemingly excessive religious rituals?

THE GAME

In the model, there are two players. One, the sender (the peacock), can send a costly signal (he can grow a long tail). The other, the receiver (the peahen), observes the signal if it is sent and then chooses how to treat the sender, namely whether to accept him (to mate with him) or reject him.

The sender might be more or less desirable as a partner, which we model by saying he can be of one of two types: a high type (a peacock with good genes) or a low type (one with bad genes). The receiver doesn't know the sender's type (it's not like the peahen can send a saliva sample to 23andMe). All she knows is whether he sent a signal (how long his tail is).

The next part is key. The signal is assumed to be costly for the sender to send, regardless of his type. However, it is less costly for the high type. The interpretation of this is that all peacocks have a harder time evading predators and avoiding parasites when they have long tails, but ones that are otherwise physically fit have an easier time and so don't get caught by predators or succumb to disease. If this were humans and fancy watches we were talking about, the idea would be that fancy watches are a waste of dough for everyone, but richer people can still manage to pay for food and shelter even without that dough.

Thus, the sender's decision is whether to send the costly signal or not (grow a long or short tail), depending on his type (high or low). He can grow a long tail regardless of his type, grow it only if he's a high type, only if he's a low type, or never. The receiver's decision is whether to accept the sender if he sends the signal, as well as whether to accept him if he doesn't. The receiver can accept all senders, only those with long tails, only those with short ones, or neither.

The key Nash equilibrium of this game is as follows. The sender sends the costly signal if and only if he is the high type, and the receiver accepts a sender if and only if he has sent the signal. That is, only fit peacocks grow long tails, and peahens only mate with peacocks who have long tails—and they do this even though the tails make their beaus worse at evading predators and more prone to disease.

To verify that this is a Nash equilibrium, we do the same thing we always do: check all possible deviations.

- Let's start with the fit peacock. Right now, he's paying the cost of the tail but receiving the benefits of being accepted. Should he deviate and stop sending the signal? Not so long as the cost of growing a long tail is less than the benefit from being accepted.
- What about sickly, low-type peacocks? Should he deviate to growing a long tail? That would require paying the cost of the tail, but it'd mean he might finally find a partner. Sounds promising, but if the tail leads to certain death in the jaws of the fox, he'll pass.
- And what about the peahen? She can deviate in two ways. She can reject peacocks with long tails, and she can accept peacocks without them. The former would mean foregoing hunky gents, and the latter means accepting sickly ones. Probably a bad idea.

Voilà! Nash equilibrium confirmed. Not only that, we also learn when it's a Nash equilibrium. Namely, so long as:

- The benefit to senders from being accepted is higher than the cost of growing a tail for the high types but not the low types
- Receivers like matching with high types but not with low types.

Among economists, this Nash equilibrium is known as a separating equilibrium because, in it, high types and low types behave differently and receivers can tell them apart, just by what they choose to do; the fit peacock need only flash his tail, and the peahen will know his type.

SOME EVIDENCE FROM SUNBIRDS

There's particularly good evidence for the costly signaling model from malachite sunbirds, with which researchers have verified not only that longer tails yield more mates but also that the tails are costly and that the costs are higher for less fit males.

To verify that longer tails yield more mates, researchers tend to take two approaches. The first is the obvious one: capture some birds, measure their tails, and follow them to see if those with longer tails have more reproductive success, which is usually measured either by how quickly they pair with a mate or how many broods they father in a season. For instance, in studies of the scarlet tufted malachite sunbird, Matthew Evans and B. J. Hatchwell found that their sunbirds' tails varied in length by over 33 percent, from just under 15 centimeters to just over 20. They also found that sunbirds with tails approaching 20 cm paired up about twice as fast as the others, and some managed to father two broods in a single season, whereas the rest only had a chance to father one.[3]

A second approach is to go in and experimentally manipulate the length of the birds' tails. This helps to isolate the causal role the birds' tail is playing in sexual attraction and ensure that females aren't being drawn to some other trait that perhaps covaries with tail length. The experiment is performed by snipping the birds' tails in half and then putting the tails back together with splints. The birds are typically randomly assigned to three treatment groups. The first has their tails shortened before being reattached. The second has them lengthened. A third group simply has their tails reattached at the same length. Then the birds are freed, and researchers wait and see who parents more broods. Those whose tails were shortened father fewer broods than those whose tails were lengthened—in Evans and Hatchwell's study of the sunbirds, roughly half as many. Tough break.[4]

But are these long tails costly? Yes. Compared to reproductive success, the tables were turned when it came to flying and hawking. Sunbirds who'd had their tails lengthened were less likely to be observed in flight, and when they hawked, the proportion of the times they successfully caught their prey fell—for some individuals, it more than halved! Meanwhile, the sunbirds who'd had their tails shortened must've suddenly felt light as a feather. They were more likely to be observed in flight and even became more fearsome hawkers, catching a slightly higher share of the bugs they went after from their perches. Their tails had been weighing them down! As for those sunbirds whose tails were reattached at the same length, they exhibited no meaningful changes.

Extravagant tails have been studied in other bird species, and other animal behaviors, such as the predator-attracting chirping behavior of nestlings (telling mom how hungry you are or how likely you are to make it out of the nest if fed well), have been shown to act as costly signals. But we think it's time to turn our focus back to *Homo sapiens*.

LUXURY GOODS

The Rolexes, mosaics, diamonds, and other luxury goods we introduced at the beginning of the chapter are a (the?) classic example of costly signals in humans. Signals of what? Not of flying or hawking ability but of wealth.[5]

This may seem self-evident, but let's nevertheless walk through how we might verify a claim like this, as the procedure will generalize.

When checking for costly signaling, we typically ask ourselves the following four questions:

Do people infer something desirable and otherwise hard to see about those who signal this way? Yes. People do infer that those with a nice wristwatch, beautiful mosaic floors, and large diamond earrings are wealthy. And wealth is something that is not, itself, easy to see; it's not like we typically have access to others' bank ledgers.

Is the signal wasteful? Yes. As we've already discussed, Rolexes are pretty robust and pretty accurate, but not as robust and accurate as a Casio G-Shock. A properly installed mosaic makes for an excellent, durable floor, but no more so than properly installed tile. Diamonds are useless outside industrial applications.

Is the signal less costly for those who send it? Yes. For a wealthy lawyer, buying a Rolex means taking on an extra case and working some late nights, whereas for most of us, buying a Rolex would mean eating nothing but rice and beans for the next three years. The price may be the same, but the cost is quite different.

Do people like the signal less if it becomes easier for others to signal with it? Once prices fall so that low types can benefit by sending the signal, receivers can no longer infer that those who send the signal are the high type. This means that costly signals have a relatively unique and surprising feature: as they become cheap and abundant, we should actually start liking them less. Is that true for luxury goods like Rolexes, mosaics, and diamonds? Yes. Nobody would be enthralled with Rolexes if you could pick them up at gas station convenience stores for forty-five dollars. Mosaics, which can now be mass manufactured for a few dollars per square foot, are no longer all the rage. What about diamonds? Synthetic ones can be cheaply produced in fancy ovens. Are they as romantic and alluring as those forged by fire, deep underground? We don't suggest you try to find out.

OTHER SIGNALS OF WEALTH (AND OCCUPATION)

So luxury goods do seem to fit the bill as costly signals of wealth. But they're far from the only costly signal you might encounter. Here are a few more. As you read on, keep in mind that, unlike in the case of conspicuous consumption, in the cases we're about to discuss, people are often less conscious that they're engaged in signaling. They just find things more pleasing that way. But from the perspective of *Hidden Games*, that doesn't matter. The sunbirds, after all, aren't consciously calculating what's Nash, either; they're just attracted to long tails.

BMI. The 2002 documentary *Fat Houses*, produced by Journeyman Pictures, begins by introducing Batya, a woman in the Nigerian city of Calabar, who is in the last months of preparation for her wedding. Batya rarely leaves her house. Her mom spends nearly the entire day cooking, while Batya rests and eats, rests and eats, with the goal of fattening herself for the big day. Her family even hires a special masseuse to massage the growing fat rolls into place to make Batya as attractive as can be to her husband-to-be.

Although a dying custom, fattening is still sometimes practiced in South Nigeria, as well as across the continent in Mauritania, a large but sparsely populated nation pinned between the Sahara and the Atlantic, where the custom is known as *leblouh*. Not long ago, fattening was still practiced across two oceans by Tahitian elites; the royal family would famously retreat to an island later purchased by Marlon Brando, where they would rest, eat, and repeat in their quest to be maximally attractive. Today, in Western cultures, most people seek to be thin, but as paintings from the nineteenth century and before reveal, in the not-too-distant past, we Westerners' tastes weren't so different from Batya's and her husband's.

The costly signaling model might help us understand why some cultures developed a taste for larger body mass indexes (BMIs) and also why that taste has changed. In places where calories were relatively scarce and costly, a taste for heavier BMIs resulted in a separating equilibrium where only those who could afford the calories fattened up. But what happened when incomes rose and the cost of a calorie tanked? Nowadays, in most places, lower-income people can afford to overeat without breaking the bank. Indeed, it's all too easy. So, high BMIs are out.

Sugar and spice and everything nice. The costly signaling model can also help us understand the dramatic change in European tastes for sugar and spice in late seventeenth-century Europe. Cookbooks and other records from the Middle Ages and early Renaissance reveal that, contrary to popular belief, Europeans of all walks of life were consummate lovers of spice. Clove, allspice, mace, cinnamon, ginger, galangal (a relative of ginger commonly encountered in Thai cuisine), black pepper, grains of paradise, and saffron were used by the handful, as was sugar—many medieval stews were quite sweet. Sugar and spice were, of course, imported and quite expensive, so their use was a luxury. Humbler families could not afford to use them as regularly and would reserve their use for special feasts, like Christmas, while merchants and the gentry ate heavily spiced food more regularly. As trade with the Indies developed, the price of sugar and spices fell, until they were within reach of even those humble families who had previously used them so sparingly. During the reign of France's Louis XIV, a new cuisine was born in the long halls of Versaille. Court chefs practically banished sugar and spice (to desserts). "Rather than infusing food with spice, they said things should taste like themselves. Meat should taste like meat, and anything you add only serves to intensify the existing flavors." From Louis's court, the fashion spread to the rest of Europe (with mixed success, as the chagrined English are quick to admit). The once costly signal of sugar and spice had ceased to be supported in equilibrium, and our tastes changed accordingly.[6]

Long (pinky) nails. In his blog *Vagabond Journey*, the author Wade Shephard writes of how it is still common to encounter men in China who grow their fingernails long.[7] The practice, which is also common in Thailand, Northeast India, and Egypt (where men typically, but not always, just grow the pinky nail long), was once even more common and has ancient roots: Egyptian men have been growing their nails long since the early Bronze Age. If you ask the men in question why they grow their nails long, a practice most of us do not find particularly attractive, they often respond that, quite to the contrary, they find it beautiful. Strange?

Not from the perspective of the costly signaling model, though long fingernails aren't a signal of wealth, at least not directly. "It is so people know that they don't do hard work," Shephard reports being told. Those who do manual labor find it prohibitively costly to grow their fingernails long, so those with more prestigious professions—teachers, politicians, doctors, and so on—separate by growing and growing to like long nails.[8]

As jobs shift away from farming, the preference for fingernails is changing. Our students are quick to tell us that, already, long fingernails are considered quite old fashioned by youth in China's cities.

Pale skin. Across East and Southeast Asia, pale skin is considered attractive, with 8.3 billion USD worth of skin-lightening products sold worldwide[9] (that's just shy of the GDP of Haiti). This is in stark contrast to our own culture, where many among us purposely sit outside in the sun or go to tanning salons to darken our skin. Then again, as medieval poems to fair ladies and impressionist paintings of women strolling in the park with parasols reveal, for the longest time, Westerners swooned for pale skin. What gives?

The costly signaling story for pale skin is the same as it was for pinky nails. Just as farmers find it costly to keep their pinky nails out of the mud, so do they find it costly to stay out of the sun. Thus, in places where the bulk of the population is still engaged in farming (or was until recently—a lag), the preference for pale skin serves to separate those with prestigious desk jobs or who are wealthy enough to have no job at all. In the West, manual laborers have, since the industrial revolution, increasingly worked indoors in factories, rather than on farms. For these workers, it is not especially costly to have pale skin. Indeed, if anything, it is costly to take vacations at sunny seaside resorts or to spend one's afternoon's lounging by the pool. Hence, the shift to preferring tan skin.

Of course, this isn't to deny the role that colorism—a remnant of racist ideals on once colonized peoples—might be playing. The world is multifaceted, and sometimes preferences have a mix of sources. Our point: costly signaling is one such source.

White dress shirts. White shirts, long—and still—a staple of the man's formal wardrobe, play a similar role in the West. "White dress shirts are the foundation to any stylish guy's wardrobe," pronounced *GQ*,[10] in an article printed just days before we wrote this, but the history of the white shirt dates back five hundred years to Tudor England. At the time, the average Englishman typically wore a jacket that covered his entire shirt. Gentlemen who, because they didn't work, were able to relatively easily keep their shirts clean began tufting out their collars and cuffs and even instructed their tailors to cut slits in their jackets.[11] This presumably allowed people to more easily see just how sparkly white their shirts were—and allowed them to separate from farmers and laborers who were unable to similarly keep their shirts clean. Over the course of centuries, jackets gradually morphed into their current form, which more clearly shows off the collar, chest, and cuffs of the sparkling white shirt underneath. Meanwhile, blue collar workers earned this name because they indeed wore blue, which effectively hid the dirt and oil stains on their shirts. (This is also why jeans are blue; indeed, the blue dye indigo, which, traditionally, was used to dye jeans, not only hides stains, it also repels them. It repels fire, too. But we digress.)

As with long fingernails and pale skin, we predict that, as jobs increasingly switch from the factory floor to the desk—or as stains become easier to remove and white easier to maintain—the sun will set on white's reign as king of shirts. In London, arguably still the capital of men's formal fashion, white shirts are still supreme but are giving ground to creative combinations of colors and patterns.

Authenticity in art. With over ten million visitors per year, the Louvre is the most popular museum in the world. Eighty percent of its visitors come to see a single item: the *Mona Lisa*. She, alone, receives more visitors than the Vatican museums. Each fan of da Vinci's mysterious maiden stands in line for one to two hours on average so that he or she can see the diminutive portrait from a distance of ten feet or so, crowded together with dozens of other eager visitors, selfie sticks at the ready. It's not the best experience. Most folks know it's not going to be the best experience. There are dozens of travel websites warning of this. But, hey, you gotta see the real thing, right?

Eh, wait. Why? In fact, if you want to get a good look at the *Mona Lisa*—to really study it in detail—you could save yourself a trip to Paris and head instead to Wikipedia (yes, Wikipedia), where you'll find various high-resolution images of the famous portrait. You can download these to your computer and pore over every pore at your leisure. Some of these images are even carefully edited to correct for the yellowing that has occurred since da Vinci's time. Pretty sweet. And certainly more informative than trying to get a peek at the original from fifteen feet away, over the heads of a Dutch family.

But, wait, those are still digital images presented on a flat screen. Even the least sentimental among us—and we *are* the least sentimental among us—can admit that's not the same thing as seeing the real *Mona Lisa*. But why not? Is there something special about seeing natural light bouncing off the painted surface? Hmmm, maybe, but we're skeptical. There are dozens of very high-quality replicas of the *Mona Lisa* available for viewing worldwide. Some require so much skill and attention to detail that they sell for upward of US $50,000. Yet, no one is queueing to see them. If it were the physical characteristics of the *Mona Lisa* that were its main attraction, then these replicas would steal at least some visitors from the grand dame of Paris. Clearly, there's something special about seeing the genuine article.

In 2011, George Newman and Paul Bloom set out to show just how much weight people place on authenticity by using a clean experimental design. The pair showed subjects two very similar paintings of a covered wooden bridge—the kind that New Englanders are just a tad too proud of—and told them that one painting was painted after the other. They randomized which of the two they said was painted first, which ensured that the physical characteristics of the paintings wouldn't drive the outcome of the experiment. In the experiment's key manipulation, they told some subjects that the paintings had been created by the same artist, implying the two were part of a series, while telling others that they'd been painted by different artists, which implied that the second painting was a forgery (*quelle horreur!*). Then, they asked them what they thought of the two paintings. When subjects were told that the paintings were painted by the same artist, they rated the second painting over three times as highly as when it was painted by different artists. Newman and Bloom's results confirm: it's not (just) a painting's physical characteristics but also its authenticity that drive its allure.[12]

Why would art aficionados put so much emphasis on the real thing? The costly signaling model suggests a simple explanation. Because authentic art is scarce, it's costly to buy or to fly to Paris to see. It's costly for everyone, but as with Rolexes, this cost is more difficult for those who aren't so wealthy. So buying or seeing the real thing is a valid way for the wealthy to separate from the less wealthy. In contrast, a forgery or a high-resolution Wikipedia image is no more costly for the poor than for the rich, so it's not a valid signal, at least not of wealth.

One prediction we can make is that our obsession with authenticity should be particularly strong for things like art, which has little intrinsic value (at an ultimate level), but weaker for things whose value comes from their utility, like a spoon or a space heater. Newman and Bloom added a treatment to their experiment to show exactly this. Instead of asking about a painting, they asked subjects about a car, under the premise that the car's value was primarily driven by its utility, namely, getting from A to B. Otherwise, the experiment was identical: they showed subjects two very similar cars and again told half that the cars were made by the same manufacturer while telling the other half that they were made by different manufacturers. Sure enough, now subjects didn't really care whether the first car had been made by the same manufacturer or another one. The two cars were about the same, so they rated them about the same. Simple as that. When signaling motives are weaker, our preference for authenticity fades.

SIGNALING OTHER ATTRIBUTES

The examples we just explored should hopefully convince you that signaling can shape all sorts of aspects of our aesthetic sense. But we have only scratched the surface. What if we expand from signaling about wealth and occupation to signaling other desirable attributes? For instance, sometimes we try to signal not just that we have money but that we *come from* money. Why do that? Here's a guess: those who come from money don't just have the goods, they have the connections with others who have the goods, as well as the wherewithal and the power to use those goods. Our next application covers such a signal. The ones that follow explore some other things we signal.

Etiquette. In Netflix's *The Crown*, Winston Churchill cajoles the queen into canceling a vacation and inviting US president Eisenhower to a banquet at the palace.[13] When she reluctantly agrees, the palace staff are sent into a frenzy of preparation. Mountains of plates, glasses, forks, knives, and spoons are brought out of the palace storerooms and piled high on rolling carts so that they can be polished, one at a time. Each chair's red velvet cushion is carefully brushed, and the dining table—as long and wide as an aircraft carrier landing strip—is cleaned by a servant who duck-walks over the surface wearing protective slippers while polishing in circular motions, so as not to miss a spot. Then Churchill falls ill, the banquet is canceled, and the tableware is returned to storage, unused.

Fans of shows like *The Crown*, *Downton Abbey*, and *Bridgerton* are often treated to glimpses of spectacular British table settings, each worthy of its own zip code. What's the deal with all that tableware? Maybe it makes sense to have a second fork for dessert, but is it really necessary to have separate ones for shrimp, salad, and the main course? To have a separate dish for our bread and three to four distinct goblets? Why bother purchasing and keeping all this tableware and learning all the ridiculous manners needed to properly use it? Why not just, you know, eat with whatever utensil is handy?

Let's use our questions from earlier to see if table etiquette is a costly signal and, if so, of what. We start by asking, What does table etiquette signal? At this stage in the chapter, you might be tempted to answer that the point of all the tableware is to signal that you can afford it. That is *part* of the point, but a small part. It's also important to know what to do with it. When people see you have good table manners, they infer you've been "brought up well" and not just that you have money.

The next question is, Is table etiquette costly? Absolutely. *Downton Abbey* had to hire a consultant to advise the cast and crew on the proper placement of the tableware and the accompanying manners. Fans of George Bernard Shaw's classic *The Pygmalion* and its more modern remakes know that Professor Higgins doesn't turn the low-born Eliza Dolittle into a duchess overnight. It takes him months of lessons, of which table manners play no small part. In the play's 1990 remake, *Pretty Woman*, the protagonist (played by Julia Roberts) is given a crash course in table manners by a friend in the hospitality business, but the lesson doesn't suffice. When the appetizer arrives instead of the salad she was expecting, she is noticed picking up a fork and counting the number of tines in effort to remember which to use.

Amusing scenes like these highlight the amount of effort that goes into learning etiquette but also how much more effort must be put in if one is to learn these rules as an adult, rather than as a child who is born into a posh family. This suggests table etiquette isn't just costly, it's costlier for those who aren't born into money. Thus, the answer to our third question—Is it less costly for those who send it?—is yes.

Fourth, and finally, we ask whether table etiquette is likely to fall out of fashion as it becomes easier for others to signal with it. There's some evidence for this. Learning proper etiquette has become easier over the course of the last century. That's partially because it's become more commonplace, and so there are more people you can learn it from, and partially because of democratizing forces like YouTube, which make it possible for anyone with an internet connection to learn the difference between a salad and a shrimp fork. Meanwhile, formal dining has somewhat fallen out of fashion. At Chicago's three-star eatery Alinea, one of the United States' most famous and most expensive eateries, the place settings are sparse and the food is occasionally even served on the table itself.

Although we've focused on table manners, rules of etiquette extend well beyond the table. For instance, after her engagement to Prince Charles, Princess Diana underwent extensive training on palace etiquette, to learn such things as: in what order to address the royals, how to address them, what one can discuss with them, in what order one should walk when in procession, and so on. Such rules—and the difficulty involved in learning them, even for those who, like Diana, have the right education—is a constant theme in the television series *The Crown*. Some of these rules do, of course, have a purpose beyond that of signaling one's upbringing (it, for instance, makes sense that the queen walks ahead of others in her entourage; see Chapter 12's discussion of symbolic gestures). But the function of some of these rules is likely signaling.

Wine connoisseurship. As any wine connoisseur will tell you, the way to develop your palate is to drink a lot of wine. So one might once again naively assume that a taste for wine is yet another signal of wealth. But this isn't enough. Connoisseurs aren't just better than the rest of us at identifying distinct flavors in wine, they're also better at talking about those flavors. To learn to talk about wine, a connoisseur can't just drink lots of wine; he has to drink lots of wine with a friend—or maybe it's an older sibling or a parent—who is a connoisseur and can train him on how to talk about what he's tasting. Wine connoisseurship therefore signals the connoisseur could not only afford all that wine but also that he has friends, or perhaps it's his parents, who were cultured enough to teach him to talk about wine.

How can we be sure that wine connoisseurship isn't *just* a costly signal of wealth? Easy. What would you think of someone who walked into a Napa vineyard tasting room and said, "I just want to taste whatever is most expensive?"[14] Such a person is clearly signaling he has cash but also signaling that he doesn't have class. The same is true for someone who buys from the best known and most expensive labels like Veuve Clicquot but shows no appreciation for lesser-known but higher-quality makers. Or someone who sticks to regions like Napa and poo-poos Paso Robles. Or someone who buys a great wine but can't say anything about how it tastes other than "good." In all these cases, the individual in question can clearly afford the wine but lacks the training—the key thing we're measuring him on.

Rhyming. Check out the following passage from the underground rapper Big L:

How **come**? You can listen to my first **album**
And tell where a lot of n***as got they whole style **from**
So what you *acting* for?
You ain't *half as* RAW, you need to *practice* more
Somebody tell this n***as something, 'fore I *crack* his JAW
You running with boys, I'm running with men
I'mma be ripping the mics until I'm a hundred and ten
Have y'all n***as like "Dammit, this n***a's done done it again"

We've grouped words that rhyme together. Those in bold rhyme with one another, then those italicized, those underlined, those in small capitals, and those shaded in gray.

Rhymes are inherently pleasing. They can enhance the rhythm of a song and sometimes make it easier for the audience to tell which word was spoken. But the rhyming schemes of songs by artists like Big L or the rappers we mentioned previously—Chance, Eminem, and MF Doom—are so complex as to suggest they're doing something more.[15] Might these complex rhymes also serve as a costly signal? If so, of what?

We think yes, of ingenuity, and we check this by finding affirmative responses to three of our four questions from earlier:

- **Do people infer something desirable and otherwise hard to see about those who signal this way?** Yes. People definitely infer that lyricists like Big L, Chance, Eminem, MF Doom, and, of course, Shakespeare are especially clever and creative. Indeed, one music website popular with music fans of all stripes, but especially rap fans, is called genius.com.
- **Is the signal wasteful?** The complex rhyming schemes are wasteful in the sense that they make it harder to say what needs to be said. Instead of choosing the first word that comes to mind, the lyricist must select groups of words that rhyme with one another and weave these groups into the story. The more complex the rhyming scheme, the more constrained the vocabulary at any point in the sentence, and the harder it is to identify groups that work together.
- **Is the signal less costly for those who send it?** Using such schemes is definitely costlier for those of us who are less practiced and talented. Imagine trying to tell someone about your day while constrained by a complex rhyming scheme such as Big L's. No, actually, don't imagine it, take a moment to try it. It would take you a while, right? Those lines above, though—Big L composed them on the fly, as part of freestyle on a day when, as he twice told the person recording the track, he was "tired."

We suspect that rhymes are just one example of how artificial constraints in art serve dual purposes: enhancing the art in some way but also providing a way for artists to signal cleverness. This might help us understand some of the constraints we encounter in the visual arts, too, like insisting on using found objects or on showing the referent from multiple angles. These constraints likely do serve the purpose people typically claim they serve. Using found objects really does call into question what art is and who gets to decide and helps us see the beauty in everyday objects. Showing an object from multiple angles really is a good way to provide additional perspectives and to make the image feel less static.

But perhaps costly signaling provides an additional reason why these constraints gained popularity. Perhaps they became popular partly because they introduced new challenges. Can you express an idea using *just* an object you found, ready-made? Can you induce an emotional reaction or convey an idea even when the referent— the thing being painted—is chopped up and the pieces jumbled together? What if you can't even refer to any real-life object at all, just basic building blocks like lines, shapes, and uniform colors? This is hard.

Why did the use of artificial constraints like these explode right when it did, at the turn of the twentieth century? We are not the first to posit this, but it might be because that's around the time producing realistic images stopped being challenging. Before that time, artists could show off their talents by experimenting with perceptual illusions, studying anatomy (da Vinci), inventing contraptions involving mirrors and lenses (Vermeer, probably), and so on—all in the service of making their paintings and sculptures as realistic as possible. When working on *The French Campaign*, Ernest Meissonier hired an army's worth of horses and models, waited until it snowed just the right amount, then had them ride out in procession over the fresh snow so that he could more accurately paint the trampled ground under Napoleon's army's feet. Such ingenious solutions showed off the artist's creativity quite well. But when the camera came along, artists had to find new challenges. This suggests that having *a* challenge is of primary importance. *Which* challenge may be of secondary importance. Of course, with each new challenge, new creative insights emerged, and our aesthetic world got richer.

Religious rules and worship. According to Chabad.org,[16] the website of Chabad-Lubavitch Hasidism, an ultra-Orthodox sect of Judaism, members upon waking up and before getting out of bed should say this brief prayer: "I thank Thee, O living and eternal King, because Thou hast graciously restored my soul to me; great is Thy faithfulness." They are then to wash their hands. When we say then, we mean *right* then:

> One is not permitted to walk four cubits [a cubit is 18–22 inches according to different authorities] before washing his hands. Before his morning hand-washing, one must not touch his mouth, nose, eyes, ears, anus; nor his clothes; nor any food; nor any place where a vein is open.

Lest you worry this not be burdensome enough, there are then thirteen additional rules governing exactly how to wash one's hands, like the following (from the Chabad.org website):

- The hands are washed in the following manner:
  a) The water-filled vessel is taken in the right hand and placed in the left.
  b) Water is poured on the right hand and then on the left hand.
  c) Repeat twice. (Altogether each hand is washed three times: right, left, right, left, right, left.)
- The water should be poured as far as the wrist except on Tisha b'Av and the day of Atonement, when the water should cover the fingers only.
- The water used for washing must not be used for any other purpose but must be spilled in a place where people do not go.

Next up are the rules on dressing and walking (yes, walking), of which there are (just) twelve. Our favorite: "A person should be careful not to walk between two dogs or swine. Also, two men should not permit a dog or swine to pass between them." That would be bad. Almost as bad as getting your wrist wet on Tisha b'Av.

So far, we've made it through just three of the eight pages of laws that govern a Chabadnik's morning routine, and we've abbreviated them quite a bit. The next five pages cover decency in the lavatory, cleanliness for prayer or study, benedictions, morning benedictions (that's not a typo—there's a page for benedictions and another for morning benedictions), and before prayer, which elucidates a few more things one is *not* to do before praying. One can, for instance, have tea or coffee but not milk, sugar, or a conversation with the neighbor before prayer begins. Oh, and it also makes quite clear that all eight pages of the morning routine should be completed before dawn since, of course, prayers should begin before dawn.

While ultra-Orthodox Jews may be extreme in their near-constant performance of religious practices, devotees to other religions around the world regularly perform a myriad of time-consuming—and sometimes painful—practices. The vegetarians of Phuket bathe in hot oil, walk on fire, and pierce themselves with sharp items. Some Shiite Muslims, Christians, and Jews practice self-flagellation. Orthodox Christians in Eastern Europe bathe in icy waters. *Etc. Etc.*

Why all these onerous rules and practices? Once one is accepted as a community member, there are all sorts of benefits. Religious people are generally more cooperative and egalitarian (toward coreligionists) and are, consequently, far more trusting and trustworthy toward each other.[17] That's been shown using lab experiments, but life experience corroborates: religious people invite each other over for Shabbat meals, offer loans or gifts to other members facing tough times, and intermarry after relatively brief courtships. This creates something of a free-rider problem: people who would like to avail themselves of the group's generosity but without sticking around long enough to have to pay that generosity back.

That's where Richard Sosis's costly signaling story comes in.[18] All those rules and practices? They act as a cost of admission for anyone who wants to be accepted as a member. Before they can be so, they have to put in the time and effort to learn the rules and perform the practices. Much of these costs are front-loaded. Ultra-Orthodox Jews must spend years studying the rules to be considered competent in performing their duties. They must buy two sets of pots and pans—one for milk and one for meat. Anyone who gets up and leaves would pay the same up-front costs but miss out on most of the benefits from being accepted as a member. Likewise for anyone who sticks around but only hangs out with community members one weekend a month at Bingo night. They would still need to invest the same amount as those who are more deeply embedded, but the benefits just wouldn't be that large. In equilibrium, those who are accepted are, in fact, the ones who can be trusted to stick around and fully embed themselves.[19] (None of this is conscious, of course! The belief in these rules and practices is genuine and deeply held.)

Along with a handful of other anthropologists, Sosis has provided a host of evidence for this costly signaling story. Here, in very brief, are some highlights.

- In places where there is more religious diversity and thus, presumably, it is easier to lose members to other religious groups, there is greater adherence to religious practices.
- Despite formal requirements that converts to Judaism be treated as equal members of the community, ultra-Orthodox (Haredi) Jews heavily discriminate against converts when it comes time to marry. Not officially but in reality. Why? Because they tend to be skeptical of the genuineness of the convert's devotion. As Sosis puts it: "It appears that those born into the Haredi community rec-

ognize that the costs of membership are too high to be paid without early indoctrination."

- The more onerous obligations a community demands of its members, the longer the community survives. There's evidence for this from communities such as Israel's kibbutzim (socialist cooperatives), as well as nineteenth-century communes in the United States.[20] Rodney Stark also argues that this is one reason why mainline religions like Catholicism, which are reducing the obligations associated with membership, are losing ground to evangelical movements, which retain high expectations of group members.[21]

- The greater the need for cooperation, the more a community demands onerous practices. After all, if there's no need for cooperation, it's not a big deal if someone doesn't stick around long enough to provide it—free riding isn't a concern. An example? Across dozens of cultures, scarring among fighting-age men is more common in communities that engage in more warfare.[22]

- The more onerous practices members of a religious group are expected to perform, the more cooperative these members tend to be. When members of different kibbutzim play public-good games, kibbutzim who have higher synagogue attendance also have members that contribute far more to the public-good games. Here we can clearly see the benefits of all those costly predawn rituals. For those who are devoted to the group—and only for them—the costs are repaid. This result has been replicated by Montserrat Soler among adherents of Candomble, a religion popular in the slums of Salvador and former slave-trading centers in Brazil.[23]

IN THIS CHAPTER, WE'VE SEEN ALL SORTS OF SEEMINGLY ODD AND wasteful tastes and practices that may be serving as useful signals of a host of different things. And that might be behind all sorts of tastes and practices you yourself have or partake in. Or spot in others and puzzle over.

Supposing you want to spot new costly signals on your own, here's our suggestion. Start by ensuring that costly signaling is at play. We gave you a handy checklist for this earlier: Does sending the signal make the sender look good? Is the signal wasteful? More wasteful for some? Does the signal go away when it becomes less wasteful? After assuring yourself that costly signaling is at play, try to uncover what, exactly, is being signaled. Questions like this might help: What kind of people find it easier or harder to send this signal? What kind of inference do others draw when they observe the signal? In what kind of contexts do people make a big deal about this signal? Does the signal itself seem optimized for conveying certain types of information?

◆

THE NEXT CHAPTER WILL FOCUS ON ONE PARTICULARLY PUZZLING WAY we signal—by making desirable signals harder to spot—and what we think these particularly puzzling signals are meant to convey: that the signaler doesn't especially care if the signal goes unnoticed by some.

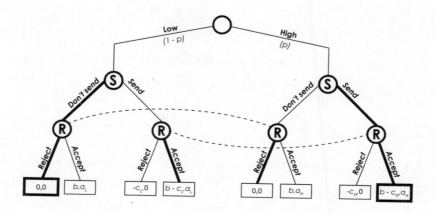

SETUP:

- The sender (S) is a high type with probability p. Otherwise he is a low type.
- The sender chooses whether to send a signal. The cost of sending the signal is $c_H > 0$ if he is a high type, and $c_L > 0$ if he is a low type.
- The receiver (R) cannot tell the sender's type, but can see the signal (that is, she cannot tell apart decision nodes connected by dotted lines). After seeing the signal, she chooses whether to accept or reject.
- The receiver only wishes to accept the sender if he is the high type. The receiver gains a_H by accepting a high type, and a_L if she accepts a low type, where $a_H > 0 > a_L$.
- The sender benefits $b > 0$ by being accepted, regardless of his type.

STRATEGY PROFILE OF INTEREST:

- The sender sends the signal if and only if he is a high type, and the receiver accepts the sender if and only if the sender has sent the signal (represented in bold). This is known as "separating" because the two types of senders behave differently.

EQUILIBRIUM CONDITION:

- The benefit of being accepted is worth the cost of sending for the high type but not the low type, i.e., $c_L \geq b \geq c_H$.

INTERPRETATION:

- In equilibrium, resources are wasted in order to signal one's type.
- Crucially, this only happens if the signal is less onerous for the more desirable type.
- Somewhat counterintuitively, if the signal becomes easier to produce, it may cease to be used.

CHAPTER 7

BURIED SIGNALS AND MODESTY

WHEN MICHAEL SPENCE FIRST DEVELOPED THE COSTLY SIGNALING game, which we presented in the last chapter, one of his primary applications was explaining why top-end employers like to hire people with fancy college degrees, even though much of what people learn at these schools won't actually help them on the job. You already know Spence's answer: these degrees help people signal a whole bunch of useful but hard-to-observe things, like their intelligence, ability to buckle down and study, social skills, family's connections, and so on.

However, if you tell Harvard students this backstory to the costly signaling game, they at first laugh nervously, but then inevitably, a student retorts that when someone asks where they go to school, they reply, "Boston," and don't clarify unless that someone insists on asking, "But which school?" If the whole point of their fancy Harvard degree is to make it easier for them to communicate to others how smart and hardworking or well connected they are—and for others to find this out—then aren't Harvard students messing the whole thing up?

In this mini chapter, we discuss why people sometimes deliberately hide, or bury, a desirable trait or impressive achievement, and why they are admired for doing so. Why do students bury that they go to Harvard, which is clearly impressive and something that others would typically want to know? Why, on a first date, might someone wealthy and successful avoid highlighting his or her high-status job or chateau in France? Again, these are things the date might well be interested in, and yet many would find it uncouth if they came up too readily in conversation. Why is it admirable when a wealthy businessman lives and dresses modestly? Wouldn't it be easier for everyone if he more readily announced his wealth and power? Why, in short, is modesty "the color of virtue"[1] when it just makes it harder to learn something rather useful? And who will be modest? When? After all, not all Harvard students hide that they go to Harvard. Some wear Harvard gear almost everywhere they go. Not all businessmen live modestly. Some live in gilded apartments overlooking Fifth Avenue.

Modesty is just one example of burying something desirable or impressive. Here are a few more that are equally puzzling.

ANONYMOUS GIVING

In season six of the TV show *Curb Your Enthusiasm*, the show's star, Larry David, donates a new wing to a local art museum. When Larry arrives at the opening, he is initially excited to see his name posted near the entrance to the new wing, but his enthusiasm is immediately extinguished when he notices that the museum's other wing was donated by "Anonymous."

"Now it looks like I just did mine for the credit," he says, worriedly, to his wife, Cheryl. Things go from bad to worse, when it turns out Anonymous is Ted Danson, Larry's longtime rival, and that everyone at the museum opening knows Ted is the anonymous donor. Ted is showered with praise. Larry is completely overshadowed and leaves the event deflated. In the cab ride home, he tells Cheryl, "Nobody told me I could be anonymous and tell people. I would have taken that option!"

Giving anonymously is puzzling for the same reason that being modest is: when a donor goes through the trouble and expense of donating, others want to know about it. It's not immediately obvious how the donor benefits from making it harder for others to do so, or why others would admire the donor for burying this useful information. Nor is it obvious who will give anonymously or when.

OVEREAGERNESS

The hottest song in 2012 was Carly Rae Jepsen's "Call Me Maybe," a "dance-pop tune . . . about hoping for a call back from a crush" that was so catchy, Justin Bieber called it "possibly the catchiest song I've ever heard."[2]

The chorus's signature line, "call me maybe," coyly highlights a puzzle: no matter how hard we fall for our crushes we go to great lengths to avoid appearing overeager. When we give a crush our number, we pretend to be nonchalant and tell the person to call us "maybe." When we get a crush's number, we wait three days to call. When a crush calls, we force ourselves to wait before calling back, and so on. We even sometimes play these games when we've met a new friend.

Why, if we're so excited about our new crush, would we want to conceal that excitement? Why isn't it better for everyone if we made our excitement clear? Then our crush would know we're interested, and there'd be no risk that he or she would accidentally think "he (or she) is just not that into me."

SHIBUI

In Japanese, the word *shibui* describes something that is seemingly commonplace at the same time that it is elegant or refined. Perhaps the glaze on a piece of pottery is muted and organic but also particularly well balanced and pleasing. Perhaps a room or garden is minimally but tastefully furnished or planted and is especially pleasant to be in or behold.

We don't have a word for it, but Westerners can also appreciate and strive for understated, or subtle virtuosity, as characterized by *shibui*. The classical pianist Grigory Sokolov, for instance, may not be a household name, but according to some critics, he is the world's greatest living pianist, lauded not just for his flawless technique but also for "transcending matters of surface display and showmanship"[3] while other, more famous pianists, like Lang Lang, Emanuel Axe, or Evgeny Kissin, "go for the show-off, virtuoso stuff."[4] Those of you who prefer pop to piano might remember how Christina Aguilera was similarly criticized for being too quick to show off her prodigious vocal abilities when she broke out on the scene in the late 1990s.

We find a similar appreciation for subtleness in the visual arts. A rather large number of paintings and sculptures can seem, to the untrained eye, as though they were created by energetic four-year-olds, when, in fact, making them required virtuosic technique and enormous know-how. Mark Rothko's large canvases of colored rectangles come to mind. After Rothko's death, museum conservators sent his canvases to high-tech labs for analysis, in hopes of illuminating the secretive artist's techniques. "Their research shows that Rothko used materials far beyond the conventional range sold for artists, modifying the properties of oil paints to achieve the flow, drying time and colours he needed."[5] Hardly the work of a four-year-old.

Why don't artists such as Rothko or Sokolov make their virtuosity easier to spot? Why does it elevate their art when they bury their virtuosity?

OUR EXPLANATION

Modesty, anonymous giving, playing it cool or hard to get, and *shibui*—what all four of these puzzling behaviors have in common is that they involve taking something desirable and, instead of shouting it from the rooftops, burying it. This seems exactly opposite of what we'd expect from the costly signaling model. In that model, the more conspicuous the signal, the better it was for the sender. So then, how can we reconcile these new puzzles with costly signaling?

The answer is actually pretty simple. We won't spell out a model this time,[6] but here's the basic idea: burying is, itself, a costly signal. A signal of what? A signal that you can afford for some people to not see some of your desirable signals.[7]

Of course, for this to be an equilibrium, as with the standard costly signaling model, burying must be costly (check) but less costly, or more worth the benefit, for more desirable senders. When might this be the case?

Here are some possibilities.

Many positive signals. The first is that some senders might have many positive attributes—a chateau, a great job, a veritable portfolio of philanthropic work, a stable family, a cute dog, and so on. Such a sender can afford to bury their positive traits because, odds are, at least one of them will be uncovered. And if it does get uncovered, the receiver can infer there must have been many more where that came from, which itself is impressive.

"You mentioned you're going to be away for spring break? Where are you headed?"

"Oh, well, my family has a home in Champagne."

"That must be so pretty. What do you do while you're there?"

"Uh, there's a charity I work with that's based not far from our place. I spend the week volunteering with them."

"Why'd your family choose Champagne?"

"Well, uh, no one knows, we've had the place for generations." Generations, eh?

In contrast, an individual with only one very particular achievement or strength would find such a strategy to be prohibitively costly. What if the conversation never takes the right turns? The receiver will walk away before you've managed to impress them.

The possibility that the sender is capable of sending so many signals that they can afford for some to remain buried may play a role when it comes to the subtle virtuosity championed by artists like

Sokolov and Rothko and prized by their fans. Perhaps these artists are so skilled that they can generate lots of buried signals, knowing some will be uncovered. Sokolov does seem to be so technically capable that he can afford to bury some of these signals. He was also sixteen when he won the most prestigious piano competition in the world and has had a following ever since. Rothko, too, was a well-trained and talented painter, who in his early years, exhibited technical proficiency in myriad styles.

Long-term relationships. Here's a second possibility. Perhaps some senders intend to interact with the receiver for a while and are willing to wait for the receiver to find out about the chateau, the job, or the charity. If the receiver doesn't uncover a particular positive trait this time because the whole conversation was spent talking about dogs at the dog park, no big deal; the receiver will uncover the trait tomorrow, next week, or in a few months. Such a sender need not blurt out that she has a chateau. She can afford to wait. And by doing so, send the signal that she is the type who is willing to wait.

Contrast that with how men in the "pickup artist" community behave. When meeting a woman, they immediately tell a practiced story they call a demonstration of higher value, or DHV, which highlights their good qualities. Unlike senders interested in long-term relationships, pickup artists need to lay it on thick and fast because if they wait, there's a risk that the receiver won't find it out in the brief amount of time they are willing to invest on this mark.

Outside options. A third possibility: perhaps the sender has other options—other suitors, lots of other interesting things to do, and so on—and isn't so desperate to match with this particular receiver. If the receiver finds out about the chateau, great, but if not, no big deal, the sender has other people to chat with and other things to do. Senders who are more desperate can't afford to do this. If they fail to impress this receiver, they'll spend the next few months alone, staring at the bottom of a glass of wine.

Devoted fans. Particularly when it comes to art, we think the following fourth possibility often comes into play: an artist might already be so well-liked and followed that, even if he buries a signal, some rabid fan somewhere will uncover it. Burying thus serves to show off that you are good enough to have rabid fans.

If Sokolov adds a subtle but difficult flourish to a simple-seeming Mozart sonata, it won't go unnoticed by the hundreds of devoted piano students who play his tracks on repeat. Perhaps the same was true for Rothko, who was already well established within the art world: he'd trained under and alongside great artists and had presented solo shows in Portland, Oregon, and in New York City before he turned his attention to creating innovative pigments.

We might therefore expect that artists will tend to bury more as they become more established. This prediction, at least, appears to bear out. Although there are probably many reasons for it, artists do tend to become more subtle as they mature. In a 2019 review, the *New York Times* lauded the historically showy and "tasteless" Lang Lang for maturing beyond obvious virtuosity. On recent tours, his repertoire had started to look more like Sokolov's: loud, fast, and passionate crowd-pleasers by Romantic composers like Chopin, Tchaikovsky, and Rachmaninoff had given way to "nuanced, delicate" performances of Mozart and Beethoven.

Specific observers. Here's a fifth and final possibility. While many artists want their art to be appreciated as widely as possible, some might be more narrowly focused on impressing a small group of fellow artists, critics, or connoisseurs. Perhaps they're not just shooting for commercial success but hoping to develop their legacy by influencing art trends and future generations of artists. Or, perhaps they really need insiders (dealers, critics) on their side to make a sale to a high-rolling collector or to arrange showcases at prominent galleries, as Rothko did with some buddies known as The Ten.

Such artists can benefit by burying the signals: if they are uncovered, their fellow artists, critics, and connoisseurs will learn that they are not after widespread fame but the kinds of things this narrower audience cares about.

Consistent with this, both Sokolov and Rothko eschew fame and popularity. Sokolov refuses to enter the recording studio or to tour in America, even though both would greatly enlarge his audience (and his pocketbook). Rothko is on the record poo-pooing the democratization of art:

> It is really one of the most serious faults which can be found with the whole conception of democracy, that its cultural function must move on the basis of the common denominator. Such a point of view indeed would make a mess of all of the values which we have developed for examining works of art. It would address one end of education in that it would consider that culture which was available to everyone, but in that achievement it would eliminate culture itself.

Contrast Rothko's quote with some from less highbrow artists, who seem quite thrilled to engage with as wide an audience as possible. When Thomas Kinkade, the best-selling painter of "bucolic and idealized scenes . . . of homey cottages and rural churches and rivers flowing gently through brilliant foliage" was accused of selling kitsch, his response was unashamed. "Everyone can identify with a fragrant garden, with the beauty of a sunset, with the quiet of nature, with a warm and cozy cottage."[8] And when Kirk Hammett, Metallica's long-haired, fast-fingered guitarist was asked what he thought of accusations that the band had sold out, he responded, "Sold out? Sure we've sold out. Last tour, we sold out almost everywhere we played." Importantly, even highbrow artists and critics typically begrudgingly admit that Kinkade and Hammett are skilled. Skilled, but also obvious.

If artists are motivated to bury because they particularly wish to impress other artists, critics, and connoisseurs, we'd predict that they'll bury in ways that artists, critics, and connoisseurs are more

likely to uncover than the typical audience member. That's almost
certainly going on. Anyone can tell when a musician's fingers move
quickly or when a landscape looks realistic or inviting, but only
those with experience know that playing slow, quiet passages deli-
cately is quite difficult or would even fathom inquiring about how
an artist makes his paints. So that's why artists might go to the
trouble of not only doing something cool but also of making it
particularly hard for most of us to notice.

The possibility that a sender mostly cares to impress specific observers may help to motivate anonymous giving. When some people
donate to a charity or a museum, they care if the average museum-
goer associates their name with philanthropy. Other donors are
part of a tight-knit community and care more about developing
their relationships with those in this community. Ted Danson, for
instance, seemed to mostly want to impress Cheryl, Larry's soon-
to-be ex-wife, who he knew would be at the museum-wing unveil-
ing and would be likely to discover the identity of Mr. Anonymous
via whispers from other attendees. (In fact, Ted didn't leave it to
chance, and told her himself.)

The Boston Brahmins, for instance, were a group of New England families who, for centuries, were the primary supporters of
Boston charities and institutions, such as the Museum of Fine Arts
and the Boston Symphony Orchestra. The Brahmins were famous
for being highly secretive, nearly always giving anonymously. In a
relatively recent article on the Brahmins, few agreed to be inter-
viewed. When the frustrated author finally asked why, she was politely told that "a Brahmin should only be in the newspaper when
he is born, when he marries, and when he dies." The families were
also famously tightly knit; there's an old ditty that goes: "And this
is good old Boston, the home of the bean and the cod, where the
Lowells talk only to Cabots, and the Cabots talk only to God."

The Brahmins' insistence on remaining anonymous meant
that the average Boston museumgoer and concert attendee did
not know about their generosity, but as Ted Danson so amusingly

demonstrated, that doesn't mean that this information couldn't be leaked to close acquaintances. Losing out on accolades from the average Bostonite was not so costly, if Brahmins like the Lowells indeed mostly cared to impress other Brahmins like the Cabots, who were bound to find out.

◆

BURIED SIGNALS, JUST LIKE THE OTHER COSTLY SIGNALS WE DISCUSSED in the last chapter, have a particular form that hints at what information is being conveyed and who benefits from conveying it. Whether it's that you have devoted fans, many outside options, or are only interested in long-term relationships, burying your signal is the perfect way to convey that particular type of information.

In the next chapter, we will consider an entirely different kind of signal—evidence.

CHAPTER 8

EVIDENCE GAMES AND SPIN

IN THIS CHAPTER, WE ARE GOING TO START BY CHARACTERIZING THREE ways people spin evidence: biased revelation, biased search, and confirmatory testing. We will then present three related models meant to help us understand when to expect these forms of spin and why they persist.

BIASED REVELATION

As anyone who spends any time on Instagram knows, the first rule of Instagram is to only post your best photos on Instagram. A casual browse of your feed on the world's third-largest social network reveals that 'grammers follow this maxim to a tee. The network is full of exotic travel destinations, scrumptious food porn, and our most flattering selfies, with nary a messy bedroom, burnt dinner, or bald spot in sight (in the rare case when an unflattering photo is posted, it is almost always posted for humorous or ironic effect). There are even apps that use machine learning to help you pick your most flattering selfies.

When people spin by only revealing the good and withholding the bad, we call this biased revelation.

Biased revelation is certainly not relegated to Instagram. When we introduce our friends and colleagues, we highlight their most redeeming qualities (Ryan bakes *amazing* bread) and leave out their less-endearing ones (wouldn't touch it 'cause that bro *never* washes his hands after he pees). When it comes time to make an online dating profile, "make sure to emphasize your best features,"[1] the how-tos proclaim, openly advocating biased revelation. When it comes time to write a résumé, the how-tos are just as blatant, telling eager job seekers to "focus on your achievements" and that "what you exclude is just as important as what you include."[2]

Political news is another domain in which biased revelation is rife. In late 2018, the US cable news networks were awash with coverage of migrant caravans approaching the US border. On MSNBC and CNN, coverage of the migrants focused on families fleeing from gang violence and political repression. Coverage on Fox focused on the young men who would, supposedly, wreak havoc on our country. In fact, both families and young men were present in the caravan, but the channels were able to selectively present the subset that best fit their preferred narrative and conveniently leave out the other half.[3]

Around the same time that the caravans were dominating the news cycle, a Florida man named Cesar Sayoc Jr. went on a terror spree, mailing homemade bombs to celebrities and government officials. The pattern quickly became clear. The targets (former vice president and future president Joseph Biden, Senator Cory Booker, former secretary of state Hillary Clinton, CNN, Robert De Niro, Senator Kamala Harris, former president Barack Obama, George Soros) were prominent Democrats or outspoken critics of Donald Trump. Sayoc was no criminal mastermind, and within a few days, authorities identified him and took him into custody. A few days later, they also took his car—a white van plastered with pro-Trump stickers. When Fox News reported on this, Twitter lit up with accusations that Fox had blurred out the stickers to withhold from their viewers evidence that Sayoc was a Trump supporter.[4]

Word counts gleaned from transcripts of the networks' coverage can help reveal the extent of biased revelation on cable news.

During the first half of Trump's presidency, the words *Mueller,*
Flynn, Putin, Kremlin, Moscow, Russia, and *obstruct* were spoken
much more frequently on CNN and MSNBC, which intensely covered the investigation into whether the Trump campaign had col-
luded with Russia, than on Fox News, where the frequent billing
of words like *uranium, Benghazi,* and *server* suggested a continued
focus on Hillary Clinton and her scandals.[5]

Like the networks that cover them, politicians are masters of
biased revelation. Take the State of the Union address—any State
of the Union address. In 2015, President Obama proudly told Congress in his opening remarks that "our combat mission in Afghanistan is over," but left out the fact that there were still thousands of
American troops in Afghanistan and no plan to bring them home
(President Biden finally withdrew the troops in 2021). Ten years
before, President Bush opened his 2005 State of the Union with an
equally rosy picture of America's foreign entanglements:

> As a new Congress gathers, all of us in the elected branches of
> Government share a great privilege: We've been placed in office by
> the votes of the people we serve. And tonight that is a privilege we
> share with newly elected leaders of Afghanistan, the Palestinian
> Territories, Ukraine, and a free and sovereign Iraq.

Bush, too, left out a lot of pretty important details, don't you
think? Well, at least we've found one thing Democrats and Republicans can agree upon: that it's better to leave the bad news out.

Listen to any presentation by a corporate executive, and you
will also hear biased revelation. For a reliable source of examples,
check out transcripts of earnings calls—quarterly calls in which
executives report their company's earnings to investors. These are
available online on investing websites like Seeking Alpha or The
Motley Fool. Click on a call transcript of your choosing, and it'll be
replete with words like *confident* and *excited* and phrases like *best*
ever and *strong fundamentals.* Words like *worried, disappointed,*

and *underperformed* don't make the cut. Here's Apple CEO Tim Cook on the company's July 30, 2019, earnings call:

> This was our biggest June quarter ever—driven by all-time record revenue from Services, accelerating growth from Wearables, strong performance from iPad and Mac and significant improvement in iPhone trends. These results are promising across all our geographic segments, and we're confident about what's ahead. The balance of calendar 2019 will be an exciting period, with major launches on all of our platforms, new services and several new products.

No mention of the fact that revenues from Apple's most important product, the iPhone, dropped 12 percent, which was more than most analysts expected. And yes, this was technically Apple's "biggest June quarter ever" but by just 1 percent. Funny how Mr. Cook left that detail out.[6]

Biased revelation is so common, it's second nature. And it's not like we don't know it's going on. We all know people post their most flattering content on social media or their most impressive achievements in their résumés. We all know that CNN, MSNBC, and Fox are selective about what they report. We can all tell that politicians are spinning (as the old joke goes, we just need to check if their mouth is moving). Companies and their execs? We call it PR.

BIASED SEARCH

Another common way in which we spin is to search extensively for favorable evidence, while making sure to cut off any search for unfavorable evidence.

Hubris, by Michael Isikoff and David Corn, is an account of how the Bush administration built support for its war in Iraq. In the book's opening, Isikoff and Corn describe how the administration, primarily led by Vice President Dick Cheney, worked with the CIA to provide evidence that Saddam Hussein had amassed weapons of mass destruction (WMDs) and sponsored terrorism. Although officially billed as an intelligence-gathering exercise, interviews and documents reveal that "Bush and his aides were looking for intelligence not to guide their policy on Iraq but to market it. The intelligence would be the basis not for launching a war but for selling it."

The administration was dogged in its pursuit of evidence that would support the invasion of Iraq:

> Cheney fixated on . . . Baghdad's ties to terrorists, especially the allegations of a connection between Saddam and al-Qaeda. The agency would write up answers to the vice president's repeated questions and send them to his office, often reporting that there was little to substantiate Cheney's darkest suspicions of an operational alliance between Saddam and Osama bin Laden. But Cheney and his hard-nosed chief of staff, I. Lewis Libby (who went by the nickname of Scooter), were never satisfied and continually asked for more. "It was like they were hoping we'd find something buried in the files or come back with a different answer," Michael Sulick, deputy chief of the CIA's Directorate of Operations, later said.

But, when it came to unsupportive evidence, the administration ignored it and suppressed further searches:

> There was no doubt. Information from intelligence analysts or other experts in or out of government that contradicted or undermined the operating assumptions of the get-Saddam crowd was ignored or belittled.

It's a tragic tale partly because the inside account of the intelligence mess is replete with episodes in which intelligence analysts and government officials actually made the correct calls about Iraq's weapons, Baghdad's supposed ties to al-Qaeda, and the difficulties that a war would bring. But they either did not prevail in internal bureaucratic scuffles or were disregarded by a White House committed to (or hell-bent on) war against Saddam.

Cheney and Bush are hardly the first politicians who, in the service of state propaganda, engaged in an intensive search for supportive evidence, while minimally searching for unfavorable evidence. Five hundred years earlier, when England's King Henry VIII made up his mind to marry Anne Boleyn, precipitating England's break with the Catholic Church and the English reformation, Henry VIII's trusty chief minister, Thomas Cromwell, recruited a small army of England's most talented theologians to help find evidence supporting the king's claim that he, and not the pope, was the head of England's church. These theologians traveled the kingdom for helpful tidbits gleaned from the dusty texts in monastery libraries and assembled them into long tomes that, conveniently, made no mention of the overwhelming majority of texts that advocated for, or took as a given, the pope's ultimate authority.

More recently, the battle over Brett Kavanaugh's nomination to the US Supreme Court provided us with vivid examples of biased search. When Kavanaugh was accused of sexual assault, Republicans in Congress investigated the living daylights out of Kavanaugh's accuser, going so far as to interview men she dated in grad school, some twenty-five years prior, in hopes of digging up something incriminating.[7] Meanwhile, the FBI was instructed to investigate the allegations against Kavanaugh but not too hard—the scope of the investigation was heavily restricted by Trump,[8] who clearly wished that the investigation turn up no new incriminating details about Kavanaugh himself.

Here's another example of biased search that most readers who grew up in the United States will recognize. In grade school, American students are taught to write essays using the five-paragraph method. Here's the Wikipedia entry, describing the function of each of the five-paragraph essay's paragraphs:

The introduction serves to inform the reader of the basic premises, and then to state the author's thesis, or central idea. . . . In each of the three body paragraphs one or more identified (evidence/fact/ etc.) that supports the thesis statement is discussed. And in the conclusion . . . everything is analyzed and summed up.[9]

Choose a position, state it in paragraph one, and then comb your readings for supportive evidence with which to fill the next three paragraphs. Think you can just as easily dig up evidence for the opposite side? Don't worry about it—that's not part of the assignment. The five-paragraph essay is teacher-sanctioned spin.

CONFIRMATORY TESTING

In addition to biased search and revelation, we also spin by generating evidence in a rather biased manner: the evidence is likely to come out looking supportive, regardless of the truth.

In 2015, the Climate Reality Project, the nonprofit founded by Al Gore to promote climate change education, published an article on its website revealing common tricks climate deniers use to spin the facts about climate change. One of the tricks that they highlight is cherry-picking a particularly hot year, for instance 1998, and then comparing subsequent years to it so that, lo and behold, it appears Earth is not warming. Another common trick is to gleefully point out particularly cold days or a cold spell as "evidence" that Earth is not warming.[10] This, obviously, ignores both that the weather is variable and that while average temperatures are expected to increase, so is the variance, which means extreme cold is likely to become more common.

Yes, that one trend climate deniers cherry-picked is consistent with their story that Earth is not warming. But is checking whether there exists a start year that yields such a trend the best way to evaluate whether Earth is warming? Obviously not. And, as with biased revelation and biased search, the puzzle is: Why will a climate denier gleefully tweet an inane climate-denying tweet when the next cold spell hits, irrespective of the fact that we all know doing so is hardly a fair-and-balanced review of climate stats?

Cherry-picking is just one way we generate evidence in a confirmatory fashion. Trump's 2020 reelection campaign illustrated another in a survey where they asked: "How would you rate President Trump's job performance so far?" Respondents could choose from the following multiple-choice options: great, good, okay, and other.[11] When such survey results are revealed, they are supportive, yes. But could they have turned out otherwise? No. And do they teach us much? Also no. Another glaring example: after British prime minister Tony Blair saw the rather thin evidence on Saddam Hussein's WMD program, he proposed "orchestrating a scenario in which Saddam would refuse new WMD inspectors"—orchestrated so that Saddam would refuse, irrespective of whether he actually had WMDs.

You know who is surprisingly guilty of generating confirmatory evidence? Scientists. One way we do this is by running experiments that are consistent with our favored theory but also consistent with the alternative. Another way: since the early 2010s, the social sciences have been racked by controversy. A small group of statistical gurus have taken it upon themselves to show that many famous results don't hold water—that they are statistical flukes and that when you run the same experiment again, they don't replicate (the controversy is often called the replication crisis). Yikes. (A few years earlier, medical research was racked by the same controversy. Double yikes.) How could this be? Researchers have all sorts of tricks up their sleeves to keep searching until they find the results they are looking for. They can run several variants of an experiment but only report those experiments that showed a result. They can run several different statistical tests and only report the tests that yielded the best results. They can stop an experiment as soon as its effect becomes statistically significant. You spend enough time and money on this, you will get the result you want. Albeit one that won't replicate and doesn't teach us much. And therein lies the problem.[12]

BIASED REVELATION, BIASED SEARCH, AND CONFIRMATORY TESTING are three of the many ways we spin. Now, we'll analyze how they are sustained in equilibrium.

THE GAME THEORY

To address these questions, we're going to develop three simple games—really, variants on a single game. The games are designed to characterize how someone (the sender) behaves when his goal is to persuade someone else (the receiver) to hold certain beliefs and he has certain tools he can use to influence those beliefs. His key tool is the control he has over evidence, which is why these games are called evidence games.

Before getting into the specific models, we are going to spend a bit of time walking through what we mean when we use words like *evidence* and *persuasion*.

States, Priors, Posteriors, and Persuasion

Let's start by defining states. The state is just the thing the sender wants to persuade the receiver about. A job candidate might wish to persuade the recruiter that she is qualified for the job. A defendant might wish to persuade the jury he is not the murderer. In both these cases—and all that we will consider in this chapter—there are just two states, which we will call high and low.

Whenever we use states, we also need to tell you the probability that they occur. We'll assume the high state occurs with some probability, p. That's the *prior* probability—what everyone believes before evidence has entered the picture. For instance, it might be the case that .3 of job candidates are qualified. Or that .65 of defendants in murder trials are guilty.

The receiver does not know the state. She starts off with the aforementioned prior and will eventually update this prior into a posterior belief based on the evidence she sees and what she expects.[13] It will take us some time to explain how the receiver forms this posterior. What's important right now is that this posterior determines the sender's payoffs.

The sender's payoffs don't, however, depend on whether the receiver's beliefs are accurate. They also don't depend on the evidence or even on the state. Rather, the sender wants to maximize the receiver's posterior that the state is high, regardless of the actual state. That is, regardless of whether the job candidate is qualified, the candidate would prefer that the recruiter believe she is qualified, and regardless of whether the defendant is guilty of murder, he would prefer that the jury believe he's not a murderer. When we say persuasion, this is what we mean—that the sender's payoffs are increasing in the receiver's belief that the state is high.

Evidence

Next, let's talk about evidence. The key thing that evidence does is provide information about the underlying state. A certificate indicating the applicant was awarded valedictorian is evidence that she is a good job candidate. A murder weapon in the defendant's closet is evidence that he is the murderer.

However, there are some additional features that we'd like evidence to have. First, a specific piece of evidence, like a certificate indicating someone is a valedictorian, either exists or it doesn't exist. The job applicant either has such a certificate or she doesn't. A murder weapon is either on the defendant's property or it isn't. Second, if the evidence exists, we can easily verify that it exists. The job applicant can show it to the job interviewer. She can scan a certificate and email it or tell the interviewer where she got it from; the interviewer can call the school and verify that she was valedictorian. Third, we cannot easily verify that evidence does not exist. We cannot, for instance, easily prove that there is no murder weapon, stashed away somewhere. Fourth, the applicant cannot easily fabricate evidence—or get away with fabricating it. Yes, she can forge a certificate showing she was a valedictorian. But that would require special equipment. And maybe the interviewer just calls the school and asks it to check the records. And then the interviewer might call the papers, or maybe even the cops. So, for the purposes of our models, we will just assume it's impossible to fabricate evidence.

When it comes to modeling evidence, the first thing we do is assign the probability that evidence exists in each state. The probability that evidence exists in the high state is q^h, and the probability that evidence exists in the low state is q^l. So if good job candidates are awarded valedictorian 6 percent of the time, and bad job candidates are awarded valedictorian .1 percent of the time, $q^h = .06$ and $q^l = .001$.

Assuming q^h and q^l are known (later we will alter this assumption) and assuming one always sees evidence if it exists (we will also alter this assumption), then it is pretty straightforward to update one's beliefs about the state after one has observed the evidence. This requires Bayes' rule, which is just an equation that tells us the probability of the high state given the evidence observed. When you obtain evidence, Bayes' rule says that probability is equal to $pq^h / (pq^h + (1 - p)q^l)$, and when you don't, it's $p(1 - q^h) / (p(1 - q^h) + (1 - p)(1 - q^l))$. In our example, this would mean that if you saw a valedictorian certificate, then you would believe the candidate was qualified with a probability of .96 (which reflects the fact that most valedictorians are qualified). Whereas, if you saw no certificate, then you would believe the candidate was qualified with a probability of .28, which is a bit lower than, but not that different from, our prior of .3 (which makes sense since there are still plenty of qualified job candidates who aren't valedictorians).[14]

Characteristics of Evidence

We'd like to highlight three characteristics of evidence that'll prove important in our analyses.

When evidence causes you to increase your belief in the state, we say the evidence is supportive or favorable. The valedictorian certificate is an example of such evidence because seeing it makes an interviewer more likely to believe the receiver is a qualified candidate (when she sees the evidence, her posterior is .96, whereas her prior is just .3). Bayes' rule teaches us that evidence is supportive if it is more likely to be generated in the high state than the low state—that is, if $q^h > q^l$. Notice that evidence can be supportive even if q^h is quite small, as long as it is greater than q^l. Indeed, that's what we just saw in our example, where q^h was just .06, but was still larger than q^l, which was .001. That is, even though most qualified candidates are not valedictorian (after all, so few people are), being a valedictorian would still increase the interviewer's posteriors that the candidate is qualified.

Evidence may be likely or common if both q^h and q^l are high and uncommon if both are low. This is independent of whether the evidence is supportive or not. In our example, most job candidates aren't valedictorians. Indeed, only .3*6% + .7*.1% = 1.87% of all job candidates are valedictorians. This form of evidence is pretty uncommon (but highly supportive!).

Evidence can also be said to be diagnostic or informative if seeing it would alter your posterior by a lot. The larger the difference between q^h and q^l, the more the posterior changes and, thus, the more diagnostic the evidence. In the most extreme case, where $q^h = 1$ and $q^l = 0$ (or where $q^l = 1$ and $q^h = 0$), you are fully informed after observing the evidence and know exactly what the state is, regardless of what you believed beforehand. This is as much as we could possibly expect posteriors to update. The other extreme is the case where $q^h = q^l$. In this case, the posterior when evidence is observed is exactly equal to the prior, p, so one learns nothing from seeing evidence. Notice, again, that whether evidence is diagnostic is independent of whether it is likely or supportive.

Coming up: our first model, in which we are going to investigate what kind of evidence the sender will choose to share. Later, we will ask what kind of evidence the sender will expend effort searching for and finally what kind of evidence to try to generate.

MODEL 1: REVELATION

In this model, we will assume the sender observes the evidence if it exists and gets to decide whether to reveal it to the receiver. The receiver can only see the evidence if the sender chooses to share it with her. When she doesn't see the evidence, she can't tell if this is because the evidence doesn't exist or if the sender just chose not to share it. To keep things simple, we will assume the sender makes his decision about a single piece of evidence at a time, with known values of q^h and q^l. Although this setup is so simple, it nicely captures some of the key features of evidence that we highlighted earlier. In particular, notice that if the sender obtains evidence, he can verify this by passing the evidence on. But, on the other hand, the sender cannot fabricate evidence or prove that he hasn't obtained it.

The question we are focused on is: For a given value of q^h and q^l, will the sender reveal the evidence or withhold it? As you can tell, the sender's choice of what to reveal may be a function of q^h and q^l (as well as p). In theory, the sender could do all sorts of things, like only share evidence if it is uncommon or if it is highly diagnostic.[15] Or he could share the evidence whenever there are low enough priors.[16] In practice, we will be particularly focused on just one strategy, in which the sender reveals the evidence if and only if $q^h > q^l$—that is, he reveals the evidence if it is supportive and withholds the evidence if it is not. Let's call this the supportive revelation strategy.

Next, let's see how the receiver forms beliefs. We're going to assume the receiver does this, taking into account what she expects the sender to do and what she sees. To see what we mean, suppose the receiver presumes the sender is using the supportive revelation strategy and that $p = .3$, $q^h = .06$, and $q^l = .001$, like in our valedictorian example from earlier.

When the evidence is revealed to the receiver, calculating her posteriors is straightforward—it's just the calculation from earlier, which yields .96. When she doesn't see the evidence, she presumes the sender didn't get it (since it's supportive, so would have been revealed had he gotten it). We did this calculation earlier, too. The resulting posterior is .27.

What if we choose another example, in which the evidence isn't supportive, say $p = .3$, but $q^h = .001$ and $q^l = .06$. Well, now the receiver doesn't expect the sender to reveal evidence, so if she doesn't see any, she makes nothing of it. Literally nothing—her posterior is the same as her prior. It's just $p = .3$.

If the receiver did see evidence, this would be a surprise, but she would still update her posterior (to .007—yikes!).

In case it's helpful, here's a timeline summarizing all the mechanics of the game:

1. The state is determined. It is high with probability p. Then the evidence is determined to exist or not. It exists with probability q^h if the state is high and q^l if the state is low.

2. The sender observes whether the evidence exists. If it does, he decides whether to reveal or withhold it. This decision may depend on p, q^h, or q^l.

3. If the evidence exists and the sender reveals it, then the receiver sees it. Otherwise, the receiver sees nothing and can't tell whether this is because the evidence didn't exist, or if it exists but was withheld. The receiver updates her beliefs about the state based on what she sees and how she expected the sender to behave.

4. The sender receives payoffs based on the receiver's beliefs (which are strictly increasing in the receiver's beliefs that

the state is high, but otherwise independent of the state and whether the evidence exists or was shared).

Here's our main result: the supportive revelation strategy is the only equilibrium of this game; that is, in equilibrium, the sender reveals evidence if and only if it is supportive (if and only if $q^h > q^l$).[17] The receiver's beliefs are formed as described above, and in particular, if she doesn't see unsupportive evidence, she doesn't make so much of this, since she knows the sender wouldn't reveal it anyhow.[18]

To see why this is Nash equilibrium, we can use the standard technique: check whether the sender can benefit by deviating. Since the sender can't do anything when he doesn't obtain evidence, we just need to check two possible deviations: withholding evidence that is supportive or revealing evidence that isn't favorable.

Withholding favorable evidence like a valedictorian certificate isn't especially helpful. The receiver expects the sender to reveal such evidence. When she isn't shown it, she assumes the sender didn't get it and adjusts her posterior downward. In our example from earlier, with $p = .3$, $q^h = .06$, and $q^l = .001$, deviating from revealing favorable evidence to withholding favorable evidence would reduce the receiver's posterior from .96 to .28, a definite drop. In general, whenever $q^h > q^l$, it will be true that this deviation leads to lower posteriors and, thus, lower payoffs.

What about deviating to revealing evidence that is unsupportive? Right now, before any deviation, the receiver doesn't expect the sender to reveal such evidence, and so her posterior will remain the same as her prior when no such evidence is shown. But if the unsupportive evidence is revealed, she'll be surprised but nevertheless incorporate the evidence into her beliefs. Since the evidence is unsupportive, this will cause her beliefs to drop. For instance, when $p = .3$, $q^h = .001$, and $q^l = .06$, her beliefs would fall from .3 to .007. Mentioning the murder weapon in the closet isn't a great way to get the jury on your side.

Notice that in equilibrium, the receiver knows the sender is selectively withholding evidence and interprets what he shows accordingly. Still the sender is doing as best he can.

That's it for the first model. Next, let's show the sender will likewise only search for supportive evidence.

MODEL 2: SEARCH

In the last game, we assumed the sender automatically observes the evidence if it exists. In this game, we'll explore what happens when the sender needs to expend effort to find the evidence. The crucial assumption in this model is: the receiver can't observe how much effort the sender has expended on searching. This means that if she does not see evidence, she cannot tell if this is because the evidence does not exist, or the sender did not search hard for it.

To model this, we start as we did before. There are again two states (high and low, the state is high with probability p), which the players do not directly observe. Again, evidence isn't guaranteed to exist. It exists with probability q^h in the high state and q^l in the low state.

Whether the sender obtains the evidence depends on whether it exists, as well as how hard he searches for it. For simplicity, we will allow just two levels of search: searching minimally and searching maximally. If the evidence exists, the sender's chance of obtaining it is f_{min} when he searches minimally and f_{max} when he searches maximally, where $1 \geq f_{max} > f_{min} \geq 0$. This probably goes without saying, but the sender has no chance of finding the evidence if it doesn't exist, regardless of how hard he searches.

The sender must pay to search harder for evidence. It costs him 0 to search minimally and $c > 0$ to search maximally; c might, for instance, represent the hours of work spent by CIA agents to dig through satellite photos and telephone transcripts or interview sources.

Crucially, the receiver can't directly observe how hard the sender searched. We the people did not know how many hours of work the CIA agents spent digging for evidence.

From this point, everything proceeds pretty much as it did in model 1: if the sender obtains evidence, he chooses whether to share it with the receiver. Then the receiver forms posteriors based on what she saw and how she expected the sender to behave. Finally, the sender gets payoffs that are increasing in the receiver's posterior.

As in the last model, the sender's choices can depend on q^b, q^l, and p, and he can, in theory at least, do all sorts of things, like search maximally only for evidence that is particularly diagnostic then reveal all evidence he finds, or search only when priors are particularly low but then never reveal. In practice, though, we can once again narrow our attention, this time to the strategy in which he searches for all supportive evidence maximally and reveals whenever he obtains it, while searching for all unsupportive evidence minimally and never revealing it—that is, search maximally and reveal if and only if $q^b > q^l$. We can call this strategy supportive search. As you've likely guessed, the supportive search strategy will prove to be the sole Nash equilibrium of this evidence game.

How does the receiver form beliefs when she expects the sender to be playing the supportive search strategy? Let's use the same parameters as in our example from earlier ($p = .3$, $q^b = .06$, and $q^l = .001$). We can assume that the chances of obtaining evidence when it exists are $f_{min} = .05$ when the sender searches minimally and $f_{max} = .95$ when the sender searches maximally. Since this evidence is supportive, the receiver expects the sender to search maximally for the evidence and, if he obtains it, to reveal it. If the receiver sees evidence, her posterior is the same as before: .96.[19] If the receiver doesn't see evidence, her posterior is .288.[20] This posterior takes into consideration the fact that the sender searched for evidence maximally. Had the receiver expected the sender to search minimally, her posterior would have still fallen, but just to .299; that is, when the receiver expects the sender to have searched more intensively, she shades her posterior belief more when she isn't presented with evidence.

When Hans Blix, who headed the UN's investigation into Iraq's WMD program in the lead-up to the Iraq War, argued against invading Iraq, he emphasized how intensively the UN had searched: "there were about 700 inspections, and in no case did we find weapons of mass destruction." The intent was, presumably, to cause his audience to likewise shade their posterior belief—to interpret the lack of evidence as more damning in light of the UN's maximal search.

In case it's helpful, here's the full timeline:

1. The state is determined. It is high with probability p. Then, the evidence is determined to exist or not. It exists with probability q^h if the state is high and q^l if the state is low.
2. The sender chooses whether to pay 0 to search minimally or $c > 0$ to search maximally. This decision may depend on c, p, q^h, or q^l. This decision is not observed by the receiver. If the evidence exists, the sender finds it with probability f_{min} if he searched minimally and and f_{max} if he searched maximally, with $1 \geq f_{max} > f_{min} \geq 0$.
3. If the evidence is found, the sender chooses whether to reveal it.
4. The receiver updates her beliefs taking into account whether she expects the sender to search minimally or maximally. And the sender's payoffs are assumed to be increasing in the receiver's belief that the state is high.

As we already said, the unique equilibrium of this evidence game is for the sender to play the supportive search strategy, provided it's not too costly to search. (If searching is too costly, the sender doesn't ever search.) Too costly just means that the cost c is weighed against the anticipated benefits of searching harder and raising the receiver's beliefs if evidence is found.[21]

The receiver, meanwhile, anticipates the sender's behavior, reducing her posterior more when the sender fails to find supportive evidence even after maximal search than if the sender had searched minimally, and, as before, makes nothing of the fact that she is never shown unsupportive evidence.

Nonetheless, the sender cannot benefit by deviating. Deviating to searching minimally for supportive evidence is all pain, no gain: it reduces the chance of showing the receiver supportive evidence, but the receiver continues to expect the sender to have searched maximally, and when the receiver isn't shown evidence, she forms her posterior under this assumption. Ouch. Deviating to searching more for unsupportive evidence is, of course, futile, since the sender would just be paying more to obtain evidence that he'd never want to reveal.

MODEL 3: TESTING

In our third game, we will consider the case where the sender can decide what tests to run and therefore what kind of evidence gets generated.

We do this by giving the sender some control over q^h and q^l. Specifically, we allow him to choose the value of q^h and q^l from some set of available values.[22] Since we already know the sender won't willingly test for unsupportive evidence, we can restrict our attention to tests for supportive evidence (so q^h is always above q^l). The receiver knows the set of available q^h and q^ls but cannot observe the particular pair chosen by the sender. After the sender makes his choice, the evidence is generated according to the state and the values of q^h and q^l that the sender chose. From this point, everything is the same as in the previous models: the sender again chooses whether to reveal the generated evidence to the receiver, and the receiver again updates based on what she is shown and what test she expected the sender to choose.

For instance, suppose that $p = .3$ and that there are two tests available to the sender. Both have $q^h = .06$, but one has $q^l = .01$, and the other $q^l = .001$. In this example, the more diagnostic test is the one in which $q^l = .001$, since this is the test that maximizes the difference between q^h and q^l. If the receiver expects the sender to choose the test with $q^l = .001$ then her posteriors are those we've already calculated: .96 if she sees evidence and .27 if she doesn't.

Meanwhile, the other test—the one in which $q^l = .01$—is more likely to yield evidence: the chances of obtaining evidence are the same in the high state but higher in the low state. However, the test is less diagnostic, since the difference between q^h and q^l is smaller. The receiver's posteriors reflect this lack of diagnosticity. If she expects the sender to choose this test, then her posteriors are .72 if she sees evidence and .29 if she doesn't. She doesn't take the evidence as seriously when she sees it, nor does she reduce her posterior as much when she doesn't see evidence.

To help you interpret the model, here's how it'd map to a real-world example. Imagine you are the sender, designing a survey about citizens' satisfaction with the president's job performance. You can decide which questions to include, which answers respondents can choose from, who to send the survey to, how many people to send it to, and so on. Each of these decisions influences the characteristics of the evidence by influencing q^h and q^l. To see this, let the high state be the state in which the public is satisfied with the president's job performance: q^h and q^l are interpreted as the likelihood of obtaining favorable responses to your survey in the high and low state, respectively. You can choose to send the survey only to the president's most ardent supporters. Then, even if the president is, on average, unpopular, a high proportion of respondents will say he's doing a great job. So both q^h and q^l will be high (you have chosen an undiagnostic test). Alternatively, you could ask a variety of questions, covering the president's successes and failures, give people the option to choose responses that express both pleasure and displeasure, and send the survey to a representative sample. In that case, you will likely get a favorable response if the president is in fact popular and an unfavorable response if he isn't; that is, q^h will be high but q^l will be low (a diagnostic test). Regardless of the kind of test you choose, others can see survey results and tell whether they are putatively supportive of the president, but they would have to dig deep to find the exact survey methods. They might, however, have enough experience or wherewithal to guess at the methods employed, which would affect how they interpret the results.

Here's the timeline summarizing the game:

1. The state is determined to be high or low. (With probability p it is high. The state is not known to either the sender or the receiver.)
2. The sender chooses q^h and q^l from a set of available pairs of q^h and q^l, with $q^h > q^l$. The sender can condition this choice on p, but not on the actual state. The observer can't see what the sender chooses, although does know the choice set. Evidence is determined to exist or not, with probability dictated by the state and the sender's choice of q^h and q^l.
3. The rest is as in model 1. The sender chooses whether to reveal the evidence, if it exists. The receiver forms posterior beliefs based on what she sees, and the q^h and q^l she expects the sender to choose. The sender then gets payoffs that are an increasing function of the receiver's posteriors.

In equilibrium, the sender will maximize both q^h and q^l. Technically, she will maximize the chance of obtaining evidence, $pq^h + (1 - p) q^l$, and reveal all evidence she obtains. This confirmatory testing strategy ensures supportive evidence is likely but also makes the evidence less informative.

Expecting this, the receiver knows not to take the evidence too seriously and updates her beliefs less than she might have otherwise. This is precisely what we saw in our example: when the receiver expects the sender to choose the test with $q^l = .01$, she updates her beliefs less both when she sees evidence and when she doesn't.

And, as in the previous two games, the sender cannot do any better by deviating to a more diagnostic test. If he chose the more diagnostic option— $q^l = .001$, in the above example—the receiver wouldn't know the more diagnostic option has been chosen and would still interpret the evidence as if it were generated according to the nondiagnostic test, with $q^l = .01$: her posterior, when she sees evidence, would therefore remain at .72, not the .96 it would be if she were informed about the chosen test, and her posterior, when she doesn't see evidence, wouldn't change either (there's no gain!). Meanwhile, the sender has reduced the likelihood the receiver sees evidence (all pain!).

Or, returning to our survey example: the optimal survey design from the sender's standpoint will be one that only asks supporters very leading questions, thereby maximizing the chance of a favorable result regarding the president's popularity. Others who see these results will not necessarily know the details of the survey design but will guess that it was confirmatory and so will not give it too much credence. Nevertheless, the president's staff wouldn't want to use a more diagnostic survey methodology, unless they could somehow prove they had done so, because others will keep assuming the survey was biased.

◆

ALTHOUGH THE THREE EVIDENCE GAMES WE JUST EXPLORED ARE EACH intended to address slightly different forms of spin, they have a lot in common. Let's discuss some of the common threads, and, along the way, double-check that the assumptions we've made are reasonable.

Private information. Private information is information that's not necessarily shared among all players. It lay at the core of each of our three evidence games models. In the first model, only the sender knew what evidence he had on hand. In the second, only he knew how hard he'd searched, and in the third, only he knew which test he'd chosen. In each case, this generated a temptation that the sender couldn't help but exploit. In the first model, the fact that only the sender knew whether he had obtained unsupportive evidence gave him the opportunity to withhold it. In the second and third model, the fact that only he knew how hard he searched or what test he used meant he could choose the one that maximized the chances of obtaining supportive evidence.

Of course, in all these cases the receiver anticipated that the sender would exploit his private information, but anticipating is not the same as observing. In the first model, the receiver knew there was some chance the sender had obtained the unsupportive evidence but still couldn't see whether he had actually obtained it in this particular case. In the second and third model, the receiver anticipated the sender's choices, but then if the sender tried to deviate from these, the receiver couldn't see this and thus couldn't reward the sender for doing so.

Is it realistic to assume senders have private information? We think yes. Instagrammers know what photos they've taken; their followers only see what they post. CEOs know more about their firms' performance than their investors. Cheney knew a lot more about what the CIA was up to than the rest of us. Researchers know their data and which tests they can run better than seminar attendees. In short, senders do often have more information than the rest of us about which evidence they've obtained and how they've obtained it.

The sender can get away with omitting information but not fabricating it. In our model, we also assumed the sender could get away with some things but not others. We assumed he could get away with withholding evidence, withholding how hard he searched, and withholding which test he employed. But we assumed he could not get away with fabricating evidence. If the sender can be sufficiently punished for withholding information, then he won't do so, and if the sender can get away with fabricating evidence, then he will.

Our assumptions are, admittedly, extreme. In real life, withholding information can sometimes get you in trouble, and you can sometimes get away with an outright fabrication. Our extreme examples are simply meant to capture the idea that it is usually easier to get away with some things (namely, withholding information) than others (fabricating). For starters: it's easier to get caught fabricating than withholding. For instance, to check whether a job applicant is lying about having been the valedictorian of his class, a job interviewer need only call the one school in question. To check whether the applicant has withheld something incriminating from her résumé, the job interviewer would need to call every school the applicant attended, past employers, any organization the applicant had volunteered with, and so on. Moreover, if caught, people tend to get sanctioned more for fabricating (a commission) than for withholding information (an omission). We will discuss why that's the case in a later chapter, but for now, we just want to point out that it generally does seem to be the case. On Instagram it's a big no-no to upload photos you didn't take (it is downright illegal in most cases), but nobody gets mad at you if you don't post photos of bad hair days. CEOs who fabricate, like Enron's Jeffrey Skilling and Theranos's Elizabeth Holmes, do usually lose their jobs and sometimes even end up in jail; whereas those who spin masterfully, like Steve Jobs or Elon Musk, are rarely penalized for this. If anything, they are rewarded for it. It's a similar story for academics. Harvard's Marc Hauser lost his job when it was discovered he fabricated data, and then when he tried to rehabilitate his career by publishing a book, his efforts fell flat. Meanwhile, few social scientists have faced meaningful repercussions for engaging in confirmatory testing. In all these cases, our assumption that fabrication is penalized more heavily than omission seems to reflect reality.

The receiver adjusts for the sender's behavior. In all three games, when the receiver formed beliefs about the state, she did so in a sophisticated way, accounting for the sender's shenanigans and calculating Bayes' rule as if it were no big deal.

In the first model, the receiver's adjustment for the sender's behavior was clearest when we compared how the receiver treated a lack of supportive evidence to a lack of unsupportive evidence: a job candidate failing to report being valedictorian on her résumé allows the interviewer to be pretty confident that she, in fact, was not her school's valedictorian, whereas a defendant not revealing that he is in possession of the murder weapon won't really lead a jury to change its assessment of whether he's the murderer.

Likewise in the second model. There, the adjustment happened when the receiver took into account how hard the sender had searched when presenting no evidence—discounting more for supportive evidence, for which the sender had searched quite hard. In the third model, the adjustment was in the dismissal, relatively speaking, of confirmatory evidence.

Do we real-world receivers account for the sender's behavior as required by our models? We do seem to draw negative inferences when others fail to show supportive evidence, especially when we presume they've searched extensively for it. We know that other 'grammers don't post rough patches on Instagram. We even know that if someone hasn't posted in a while, it might be wise to check on her, as she might not be doing well. We similarly expect a first date to keep some stuff to himself and don't make too much of the fact that he didn't volunteer anything damning. Investors know that if Tim Cook doesn't mention the iPhone on an earnings call, well, then, that iPhone probably hasn't had a stellar quarter. A historian or a journalist likewise doesn't simply assume that just because a source didn't say something damning that there's nothing damning there. The UN Security council did not take the Bush administration's side after Colin Powell's (in)famous speech, presumably surmising that the lack of strong evidence was rather telling, given that the world's largest, best-funded, and most sophisticated intelligence agencies had spent months searching for it. And most referees know that if a researcher is leaving out an obvious figure, table, or statistical test, well, that's probably because it doesn't look so hot.

We can, likewise, intuit when evidence is not so diagnostic. When someone spouts a climate stat that happens to begin in 1998, we can tell this wasn't a random choice and know to be skeptical.

Often, people can't tell exactly how others made their test confirmatory, but they know they probably did and to interpret the evidence with a grain of salt. If the Trump Campaign promoted the results of the survey that we mocked earlier, most of us would not know exactly how they'd generated 95 percent job satisfaction ratings but would nonetheless have intuited that they didn't ask the most diagnostic questions, or that they avoided giving the survey to anyone who might have disagreed. In short, while our games make high demands of real-world receivers, those receivers seem up to the task—not just up to it but engaging in it.

Of course, real-world receivers don't always fully take into account the sender's behavior. They might not fully adjust for the biased behavior of a sender if they are not fully aware of the sender's abilities and motives. For instance, they might underestimate how easy it is to rerun another experiment or to sic more CIA agents on the job of digging up interviews and photos that fit the desired narrative. Or they might presume the sender is actually unbiased—motivated to inform—when he is in fact motivated to persuade. Absent this information, receivers cannot properly adjust for the sender's spin. But that doesn't remove the sender's motive to spin or change the ways he will choose to spin. In fact, it just makes the effect of spin more damaging: in addition to the wrong information being collected and shared (true in our models too), the receiver is going to end up with biased beliefs (wasn't true in our models).

Here's another instance people might not be fully adjusting to the spin they are shown: when they are in on the con. For instance, when we consume political news, we *might* be the ones that the media is trying to persuade. Or we *might* be collecting our own evidence in the service of persuading others, and the media is just the means by which we gather that evidence. In such cases, we won't be discounting what the media presents to us because we are not the receivers—we are, in fact, the senders. This doesn't contradict our model; it just requires a bit of caution.

Despite the receiver's incredulity, the sender cannot benefit by deviating, and spin persists in equilibrium. Even though the receiver was aware and adjusting for the biased behavior of the sender, the sender could do no better than by continuing to behave that way. It wouldn't help to withhold supportive evidence or reveal nonsupportive evidence. Nor would it help to search or generate evidence more evenhandedly, since others would continue to judge what the sender presented as if he hadn't.

That is, if we want to know why Instagrammers, CEOs, Dick Cheneys, and academics engage in biased revelation, biased search, and confirmatory tests even though others are aware that they're doing so, well, our models suggest the answer is: it's an equilibrium for them to do so. The only equilibrium! Provided they have a motive to persuade, and there's private information over which evidence they have and how they obtained it, that is.

◆

FINALLY, LET'S SEE IF WE CAN GAIN SOME ADDITIONAL INSIGHTS USING our usual trick of breaking some of our models' assumptions.

So far, our senders wanted their receivers to believe they were a qualified job candidate or weren't a murderer. Others wanted receivers to believe that Saddam had WMDs. Yet others wanted receivers to believe that Trump was a popular president. Crucially, these senders wished their receivers would believe these things *regardless* of whether they are true. They were, in short, motivated to persuade.

Let's break this assumption. After all, there are many cases where the sender wants to inform the receiver rather than to persuade her. For instance, a sender who is thinking of setting up two friends might certainly hope that they will ultimately fall for each other but doesn't want them to believe it if it's just not going to happen. (That's a recipe for losing two friends.) Or, similarly, an intelligence agency might not be in the business of building a case for war but of helping the president and the public form an accurate opinion of a country's WMD capabilities. Or what of the many survey companies that don't work for campaigns and simply want their subscribers to get as clear a picture as possible on the public's opinion of the president? How would such senders behave in our three games? Would they continue to reveal and search for only supportive evidence? Would they continue to employ tests that maximize the chances of obtaining supportive evidence, even in the low state? No. In all three games, a sender who wishes to inform would behave quite differently than one who wishes to persuade.

In the first game, the sender would now reveal any evidence he could get his hands on, supportive and unsupportive. "Jeff is hard working, creative, and a great businessman. Last year, his net worth increased by 90 billion! He sometimes goes on and on about space and can, in general, be a bit intense. I should also say, he's not so tall and he's *totally* bald, but he works out, and I think he looks pretty good, especially for a tech bro. I can show you a picture if you'd like?"

In the second game, the sender would now search for both kinds of evidence, not just supportive. He'd devote resources not only to learning that Al Qaeda recruits come from Iraq (as the CIA discovered at the time) but also to learning whether there might be exculpatory explanations. He'd throw resources into learning whether Saddam was personally involved (we eventually learned he wasn't) and whether the recruits came from neighboring countries as often as they did from Iraq (we eventually learned that they did). Of course, this is precisely how intelligence agencies do normally behave, and it's precisely how, according to Michael Isikoff and David Corn, the CIA tried to behave until Cheney discouraged it from pursuing leads for unsupportive evidence.

And what about the third game? Remember, a sender who wished to persuade did what he could to increase the likelihood of obtaining evidence even in the low state, for instance, by only surveying the president's supporters or only offering responses of great, good, okay, and other. A sender who wishes to inform would absolutely not do this. Instead, the sender would design tests that were maximally diagnostic, which means *minimizing* the chance of obtaining supportive evidence in the low state. We already discussed how he might do so: ask a variety of questions that also cover the president's failures, give people the option to choose responses that express both pleasure and displeasure, send the survey to as many people as possible including those from the opposite party, and so on. And this is, indeed, what pollsters who aren't propagandists do.

Since biased revelation, biased search, and nondiagnostic tests show up when a sender is motivated to persuade but not inform, these behaviors can help us ascertain a sender's motives, when we are not sure. A friend who sets you up but only tells you the positives isn't looking for a good fit—they're hell-bent on there being a first date no matter what. A vice president who threatens to fire agents who search for unsupportive evidence, as Cheney is reported to have done, is not motivated to uncover the truth about Iraq's WMD program, he's engaged in a propaganda mission and has another agenda for going to war. A politician who parrots cherry-picked climate statistics is probably getting donations from Exxon. And so on. Biased revelation, biased search, and nondiagnostic tests are useful signatures of persuasion.

Next, let's break a second assumption: that the sender can verifiably pass on evidence he has obtained but not that he hasn't obtained evidence. As we already discussed, this assumption is probably usually appropriate. A CEO can easily pass on a particular stat about the firm's performance, but it would be hard to prove that there are no stats that investors should be worried about. An intelligence agency can relatively easily share the evidence it has found, but it is relatively difficult for it to prove that it has left no stone unturned in the search for evidence. And a researcher can easily present the results of a particular regression. It's much harder to prove she has run all the relevant regressions and they all point in the same direction.

Sometimes, though, it is possible for senders to verify that they haven't obtained evidence. If employers view criminal history as evidence that someone is not qualified for a job (and, unfortunately for ex-cons, they often do), then we'd expect job applicants with a criminal history to withhold this unsupportive evidence. However, job applicants with no criminal history can then just say so, and prospective employers can verify this claim by running a background check. Because of this, employers end up fully informed, even though some applicants might try to withhold unsupportive evidence. In fact, this is a special case of a now-classic result by the Nobel Laureate Paul Milgrom, who showed that when evidence is freely verifiable, then the receiver ends up fully informed, in equilibrium.[23]

Finally, sometimes it's possible to make omissions harder to hide. This is arguably what has happened in the social and medical sciences, which have adopted a norm of preregistration, in which researchers publicly announce which statistical analyses they plan to perform in advance of actually performing them. Preregistrations make it observable whether a researcher is omitting some of the informative tests he promised to run, eliminating the temptation to do so. By making omissions harder to hide, spin is reduced, and science is advanced. More of this, please.

◆

THE THREE GAMES WE PRESENTED ARE TAILOR-MADE FOR THE THREE forms of spin with which we opened this chapter: biased revelation, biased search, and nondiagnostic tests. But, of course, people spin in a host of other ways. For instance, on the dating website OkCupid, men are two inches taller than the national average and a surprising number are exactly six feet tall. Women are a solid one inch taller than the national average. Or are they? The most likely explanation—and, if you believe the blog posts and forum discussions, the one typically assumed by OkCupiders themselves—is that many OkCupiders exaggerate their height by a bit.[24]

This is a form of spin that's not captured by any of our three models we presented, since in our models, the sender gets evidence and passes it along to the receiver. But on OkCupid, height is self-reported. No evidence is required or presented.

Yet, the threads that were common to our three evidence games are present in this case, too. There's a motive to persuade (people typically prefer taller partners). There is private information (the OkCupider's height). There's also an asymmetry in what kinds of lies can be discovered and punished (small exaggerations, like omissions, are harder to discover or prove). In equilibrium, senders will lie but only in a limited way (exaggerating their height by about two inches, apparently), receivers are aware they are lying (and express their frustration on blogs and in forums), and yet senders are unable to benefit by deviating: if receivers expect the sender to exaggerate by two inches and he doesn't, they'll simply assume that he's two inches shorter than he actually is, and he'll end up with fewer dates. Moreover, ironically, his potential matches would actually have a *less* accurate estimate of his height.

The point being, while the details of the model will vary for different forms of spin, we expect the explanation to share the core common threads that we highlighted above.

◆

IN THE NEXT CHAPTER, WE WILL SEE HOW OUR EVIDENCE GAMES NOT only explain some quirks of how we persuade others but also some quirks of how we persuade ourselves—what psychologists call motivated reasoning.

Game 1 - Revelation

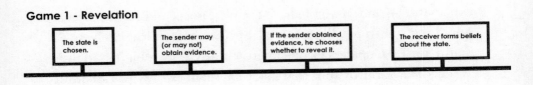

SETUP:

- The state of the world is either high or low.
- The sender has a persuasion motive, which we model by assuming his payoffs are increasing in the receiver's posterior that the state is high; the sender's payoffs do not depend on anything else.
- The sender may or may not obtain "evidence." His likelihood of doing so depends on the state: he obtains evidence with probability q^h in the high state and probability q^l in the low state. We say evidence is "supportive" if it is more likely to be obtained in the high state, i.e., $q^h > q^l$.
- The sender chooses whether to reveal the evidence, if he has obtained it.
- The receiver updates her beliefs based on q^h and q^l, whether she expects the sender to reveal, and whether the sender actually does so. Notice the receiver does not make any choice or receive any payoffs; she just updates her beliefs.

STRATEGY PROFILE OF INTEREST:

- The sender reveals evidence if and only if it's supportive.
- The receiver expects this and adjusts her beliefs accordingly. When supportive evidence isn't revealed, she presumes it wasn't obtained. When unsupportive evidence isn't revealed, however, she draws no inference about whether it was obtained, knowing it wouldn't be revealed to her.

EQUILIBRIUM CONDITIONS:

- This strategy is always the only equilibrium of this game.

INTERPRETATION:

- When will we see people presenting evidence in a biased fashion? When there is a persuasion motive and it is easier to withhold evidence than to fabricate it.
- What will count as "supportive"? Evidence that is more likely to be generated in the state that the sender wishes others to believe in.

Game 2 - Search

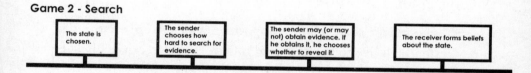

SETUP:

- Once again the state of the world is either high (with probability p) or low, the sender prefers that the receiver believe the state is high, and evidence is generated with probability q^h in the high state and q^l in the low state.
- The sender chooses whether to search minimally or maximally for evidence. If he searches minimally, he pays nothing and, if the evidence exists, he obtains it with probability f_{min}. If he searches maximally, he pays a cost $c > 0$, and, if the evidence exists, he obtains it with probability $f_{max} > f_{min}$. If he obtains evidence, he chooses whether to reveal it.
- The receiver does not observe whether the sender has searched, but does observe the evidence obtained. She forms beliefs that the state is high based on whether she observes evidence, q^h and q^l and her expectation of how hard the sender has searched.

STRATEGY PROFILE OF INTEREST:

- The sender never searches for unsupportive evidence, and always searches for supportive evidence.
- The receiver accounts for the sender's biased search. When she is shown supportive evidence, she presumes the sender searched maximally for it, and it does not change her posteriors as much as if he had searched less.

EQUILIBRIUM CONDITIONS:

- The sender searches for supportive evidence only if the c is small relative to the expected impact on the receiver's beliefs of searching harder. (The precise condition is: $c \leq (f_{max} - f_{min})$ $\Phi (\mu^1 - \mu^0)$, where $\Phi = pq^b + (1 - p)q^l$ is the unconditional probability of obtaining evidence, $\mu^1 = pq^b f_{max} / (pq^b f_{max} + (1 - p)q^l f_{max})$ is the receiver's posterior when she observes evidence, and $\mu^0 = p((1 - q^b) + q^b(1 - f_{max})) / (p((1 - q^b) + q^b(1 - f_{max})) + (1 - p)((1 - q^l) + q^l(1 - f_{max})))$ is the receiver's posterior when she doesn't observe evidence, given she expects the sender to have searched maximally. Note, for simplicity, we've assumed the sender's payoffs are linearly increasing in the receiver's posterior.)

INTERPRETATION:

- So long as searching for evidence isn't too costly, the sender searches for supportive evidence. However, he never searches for unsupportive evidence.
- When will we see people searching in such a biased fashion? Whenever there is a persuasion motive, and it is easier to hide the details of the search process than the results of the search.

Game 3 - Tests

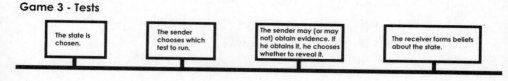

| The state is chosen. | The sender chooses which test to run. | The sender may (or may not) obtain evidence. If he obtains it, he chooses whether to reveal it. | The receiver forms beliefs about the state. |

SETUP:

- Once again the state of the world is either high or low, and the sender prefers that the receiver believe the state is high.
- Without knowing the state, the sender chooses a "test." Each test is characterized by a pair of probabilities, (q^h, q^l), taken from some set of pairs Q, where $q^h > q^l$ for every test in Q.
- We say a test is "confirmatory" if both q^h and q^l are large, so that the sender is likely to obtain evidence, regardless of the state. We say a test is "diagnostic" if $q^h \gg q^l$, so that the test is much more likely to generate evidence in the high state.
- After a test is selected, evidence is generated according to the state and the chosen test. If evidence is generated, the sender automatically obtains it, and chooses whether to reveal it to the receiver.
- The receiver does not observe the state or which test the sender has selected, but does know the set of available tests, Q. She only observes whether evidence was revealed to her. She forms beliefs that the state is high based on whether evidence was revealed to her, the test she expects the receiver to have chosen from Q, and whether she expected the sender to reveal the evidence.

STRATEGY PROFILE OF INTEREST:

- The sender chooses a test that maximizes both q^h and q^l, and so is confirmatory and not diagnostic. In particular he selects the q^h and q^l in Q that maximizes $pq^h + (1 - p)q^l$, where p is the probability the state is high. He then reveals the evidence whenever it is generated.

- The receiver expects the sender to choose confirmatory tests. Consequently, she may not update her beliefs much when she is shown evidence.

EQUILIBRIUM CONDITIONS:

- This strategy is always the only equilibrium of this game.

INTERPRETATION:

- When does the sender conduct "confirmatory tests"? When the sender has a persuasion motive and the details of the test are harder to observe than the test results.
- How do tests look different when designed by someone with a persuasion motive vs. a motive to inform? A persuader wishes to maximize the odds of a supportive result, regardless of the truth. An informer only wants supportive evidence in the case the state is high.
- As before, there is no need to assume "irrationality" on the part of the sender or receiver.

CHAPTER 9

MOTIVATED REASONING

IN THE LAST CHAPTER, WE FOCUSED ON HOW THE SENDER DISTORTS evidence when trying to persuade. However, often, Instagrammers, CEOs, and political pundits end up with beliefs that are at least as distorted as those they are trying to persuade us of. We'll explain this by adding one simple twist—which we largely owe to Robert Trivers, Bill von Hippel, and Rob Kurzban[1]—to the models presented in the last chapter: we often internalize our own spin. With this twist, along with the models from the last chapter, we will be able to explain many of our own biased beliefs as well as a chunk of the psychology literature collectively known as motivated reasoning.

Let's start by describing some of the key findings in the motivated reasoning literature. Then we will discuss internalization and how that, along with what we did in the last chapter, can explain motivated reasoning. The last few pages of the chapter will provide evidence that this is a helpful way to think about motivated reasoning—or at least a meaningful swath of it.

OVERCONFIDENCE. PEOPLE INVARIABLY THINK THEY ARE HEALTHIER, smarter, more attractive, better drivers, and so on than they actually are. Here's a classic experiment that demonstrates this.

Take a look at the following traits, and rate how well each describes you on a scale from 1 to 7 (1 = not at all characteristic of you, 7 = very characteristic of you).

Snobbish
Cooperative
Considerate
Unskilled
Compulsive
Spiteful
Melancholy
Dependable
Strict
Bold
Trustful
Eccentric
Disobedient
Resourceful
Polite
Kind

Now repeat the exercise but for the people at your school or your job. On average, how well does each trait describe them?

As you can guess, people tend to rate themselves relatively more highly for positive traits like cooperative and dependable than they do for the average person in their peer group, while rating others more highly for negative traits like unskilled and compulsive. If you're thinking to yourself that you avoided the trap, well, just so you know, most people think they're less biased at this task than the average person, too.[2]

Asymmetric updating. If it's optimal for the sender to only present supportive evidence then it's optimal for the sender to update his own beliefs in a biased fashion: to update them in response to supportive evidence but not in response to unsupportive evidence. Why internalize unsupportive evidence that one doesn't intend to present? There's no need, and worse, as Trivers and the gang argue, this introduces a chance of accidentally revealing the unsupportive evidence.

Indeed, such asymmetric updating is a mainstay of the literature on motivated reasoning. Here's a, frankly, somewhat mean study that nicely illustrates it by behavioral economists Justin Rao and David Eil.[3] In the beginning of the study, subjects were asked to rate how smart and how attractive they are. Then, they took an IQ test and were scored on how attractive they are by a panel from the opposite sex. Finally, subjects received their scores and were asked to update their assessment of how smart and how attractive they are. Like we said, kinda mean.

A fully rational person should incorporate his IQ and attractiveness scores into his self-assessment. If he receives good news, in the form of scores that are higher than his initial self-assessment, he should raise his assessment and vice versa. When Rao and Eil analyzed their experiment's data, they found that subjects indeed incorporated their assessments in response to good news. But when subjects received bad news, they barely adjusted for it. This result, which has been replicated in other studies, is often called asymmetric updating.

Asymmetric search. If you're going to believe supportive evidence but not unsupportive evidence, then what kind of evidence would you search for? Supportive? Yes. Unsupportive? Not at all. And that's exactly what people do. Here's how Dan Gilbert puts it:[4]

> When our bathroom scale delivers bad news, we hop off and then on again, just to make sure we didn't misread the display or put too much pressure on one foot. When our scale delivers good news, we smile and head for the shower. By uncritically accepting evidence when it pleases us, and insisting on more when it doesn't, we subtly tip the scales in our favor.

In a classic study, Peter Ditto and David Lopez invited subjects into the lab under the guise of studying "the relationship between psychological characteristics and physical health."[5] They asked subjects to fill out a questionnaire and then told them they'd be undergoing a simple medical test to see if they had a deficiency for the enzyme thioamine acetylase, which could lead to various pancreatic conditions later in life. Scary stuff. Fortunately, testing for thioamine acetylase was simple because a brand-new test strip had been developed to identify its presence in saliva. All subjects had to do was spit in a cup, take a test strip, and leave it in the saliva for a few seconds, and the strip would indicate if they were doomed to suffer a painful and premature death. (This was, of course, all bullshit. Thioamine acetylase deficiency was a made-up condition, and the test strips were just strips of plain old paper.)

Here's the clever part. Ditto and Lopez told half their subjects that the strip would turn green if they had the deficiency, and the other half that the strip would turn green if they didn't have the deficiency. Then, they secretly videotaped subjects as they took the test.

Subjects who were told their strip would turn green if they had the deficiency completed the test quickly. They dipped their strips into the cups, saw no change in color, and were happy to call it quits. Not so for subjects who were told their strip would turn green if they didn't have the deficiency. They took almost thirty seconds longer to complete the test. They also dipped their test strip into the cup repeatedly at three times the rate. To quote Dan Gilbert again, "Good news may travel slowly, but people are willing to wait for it to arrive."

To show that their results weren't somehow specific to medical tests, Ditto and Lopez developed a similar experiment in a totally different setting. They placed a stack of index cards in front of the subjects, blank side up. On the other side was printed an exam question that one of their classmates had supposedly completed (again, this was a fabrication). It was graded in red. Although the questions were relatively easy, the classmate had apparently gotten most of the questions wrong. Subjects began the task by answering a questionnaire that included a question about whether they liked their classmate. Then they were asked to start going through the stack of index cards. They could stop as soon as it was possible for them to make an assessment of their classmate's intelligence. Subjects who reported liking their classmate flipped over almost 50 percent more cards in hopes of finding something redeeming about their friend before finally giving up and making their assessment.

Although we've singled out Ditto and Lopez's studies, they're just intended to illustrate the more general phenomenon of asymmetric search—searching hard for supportive evidence (evidence that one is healthy or that one's friend isn't a dullard) and as little as possible for unsupportive evidence (evidence that one is sick or that one's friend is indeed a dullard).

Attitude polarization. In the late 1970s, the psychologists Charles Lord, Lee Ross, and Mark Leper invited subjects to the lab for a two-part study to demonstrate attitude polarization.[6] In the first part, they asked subjects to read this short blurb about the death penalty:

> Kroner and Phillips (1977) compared murder rates for the year before and the year after adoption of capital punishment in 14 states. In 11 of the 14 states, murder rates were lower after adoption of the death penalty. This research supports the deterrent effect of the death penalty.

After reading the blurb, subjects were asked whether they believed the death penalty helped to deter crime.

Then, in the second part of the study, the subjects were shown critiques of the study by other researchers, as well as the authors' rebuttals to those critiques. Finally, subjects were asked to rate whether the study had been well done and whether it provided convincing evidence that capital punishment helps to deter crime.

The authors found that subjects who, in part one of the study, said that they believed that capital punishment helps to deter crime also tended to say, in part two, that they believed the study was well done and convincing. In fact, they reported more strongly believing that the death penalty helped to deter crime. Meanwhile, subjects who had said they thought capital punishment doesn't deter crime saw exactly the same evidence but interpreted it differently. They latched on to the study's failings, reported that it was not well done or convincing and that they now were even more strongly opposed to the death penalty. That is, even though subjects saw exactly the same evidence, it had an opposite impact on their beliefs, depending on which side of the debate they were initially on.

Being careful experimentalists, the psychologists repeated the experiment but used a "study" whose finding didn't support capital punishment (these studies and their accompanying critiques were, in fact, made up). Again, they found that the same evidence had an opposite impact on subjects' beliefs, depending on whether they supported or opposed the death penalty.

Such attitude polarization has made more appearances in laboratory studies since this first appearance over forty years ago and also occasionally has a cameo in public discourse of current events. Some of our readers may recall a viral video involving a group of students from Covington High School wearing Donald Trump's iconic MAGA hat. The students, it seemed, were harassing a Native American man named Nathan Phillips during a protest. The video sparked a national outrage among liberals when it first emerged, but it turned out that the situation was more ambiguous than it seemed. The incident was sparked by members of the hate cult Black Hebrew Israelites, who were also present and had hurled racist slurs at the Covington boys. The boys at first did nothing, but eventually, with the permission of their chaperone, started doing their school cheers. Partway into a cheer, Phillips walked into the boys' circle and started banging his drum in a boy's face. The video captures (just) this moment of confusion, as the boys respond with smirks and verbal jabs.[7]

After these additional details emerged, conservatives were quick to defend the high schoolers and condemn the liberals who had criticized them on social media. But many liberals dug in and continued to blame the high schoolers, sometimes more vehemently than before.

◆

INTERNALIZED PERSUASION

Imagine you're at an interview trying to convince someone that you would make a good employee. Or maybe you're on a date and trying to give the impression you'd make a good romantic partner.

Now imagine that you, yourself, don't believe it. That's likely to be a problem for all sorts of reasons. You might not put effort into the interview or the date, figuring it's not going to work out so well. And you might say things that reveal why you're not a great fit for the job or for this particular relationship. Even if you watch what you say, it'll probably be obvious you're really not interested in the job or the date. You might hesitate before you speak, or it might even be apparent from your facial expressions. This is an argument that's now been made by a small crew of researchers, including the three we mentioned earlier, Bob Trivers, Bill von Hippel, and Rob Kurzban: when people believe something, others are liable to find that thing out.

This creates a strong incentive to believe that you *are* a good candidate for the job or the relationship, even if you're not, to be overconfident in how good a fit you are. But by how much should you be overconfident? Can we give some more precise predictions on how your beliefs should respond to evidence? This is where our analysis from the last chapter fits in. It suggests how people might change their beliefs in response to evidence depending on how they are expected to use that evidence.

For instance, should your beliefs reflect supportive evidence? Yes, absolutely! People expect you to present this evidence. If you don't incorporate this evidence into your own beliefs, then you can't present it, and they'll think you never got it.

What about evidence that doesn't support your case? Why risk giving away that you're aware of such evidence? Better to ignore it. And what about the fact that you searched for evidence in a biased way, or that the supportive evidence you found is not especially diagnostic? Should your beliefs reflect this? Fuhgedaboutit. Why cop to this if you don't have to? Why risk giving up the game?

If beliefs update when you get supportive evidence but not for any of the more negative stuff, then of course you'll end up overconfident. But, notice, you end up overconfident in a very particular way: your overconfidence reflects all the supportive evidence you can muster but not how hard you had to search for it, what you might be hiding, or how flimsy that evidence actually is.

Moreover, the way in which you update your beliefs will also fit the features of motivated reasoning that we encountered. You'll update them asymmetrically—responding to the good stuff and ignoring the bad stuff. And what if the evidence you encounter is ambiguous? Then you'll interpret it in the most favorable way possible and incorporate this interpretation but not the way in which you reached this interpretation. If another person with different aims sees the same ambiguous evidence you did, she'll interpret the evidence in a way that's favorable to her, and the two of you will end up being even more polarized than when you started out.

But wait, what about lies? Should you start to believe fairy tales about how you were the valedictorian or the homecoming king or queen or whatever? Probably not, since the likelihood of getting caught in a lie is pretty high, as is the cost of getting caught. Too hard to get away with it. Contrast that with what would happen if someone tries to call you out on ignoring unsupportive evidence or on generating evidence in a biased or nondiagnostic way. Then you'd have plenty of ways to weasel out of it—lots of excuses. Maybe you forgot, didn't see, didn't know what you should check, didn't understand, *etc., etc.*

SOME EVIDENCE

We will now weave together several strands of evidence in support of our suggestion that motivated beliefs are well explained by internalized persuasion. Let's start by considering whether beliefs indeed covary with their persuasion motive.

Rex Tillerson, the former CEO of the giant fossil fuel company Exxon-Mobil who subsequently served as the US secretary of state, does not deny climate change. He just says there's nothing we can do about it. "With respect to our ability to influence it," he said in January 2021, nearly half a decade into his retirement as Exxon's chief, "I think that's still an open question. Our belief in the ability to influence it is based upon some very, very complicated climate models that have very wide outcomes."[8] This belief is, of course, in conflict with the scientific consensus, which is that we can influence climate change by reducing CO_2 emissions—by burning and buying less of the oil and gas produced by his former company and his allies in Saudia Arabia and Russia. Indeed, it's in conflict with Exxon's own internal documents, which even in 1982 admitted that climate change could be mitigated by aggressively reducing fossil fuel consumption.

Mr. Tillerson holds a degree in engineering and has decades of experience reading technical and scientific documents. He knows how to evaluate risk and uncertainty—that's a necessity in the oil and gas business. He has access to cutting-edge research on climate change, and if he rang up a top climate scientist, she would drop everything to provide him with guidance. He even knows that his own company is single-handedly responsible for much of the "scientific" literature arguing against climate change. Why, then, do Mr. Tillerson's beliefs (or at least his purported beliefs; it's hard to know the exact extent these beliefs are internalized) about climate change differ so substantially from those of top scientists?

Of course, Mr. Tillerson's beliefs are no surprise once we consider his motive to persuade. As a sender in the business of

defending fossil fuels, Mr. Tillerson is likely to present very different arguments and evidence than scientists, who, though perhaps
not purely, are more motivated to inform. Arguing that cutting
down on fossil fuels is futile is precisely the argument one might
expect of such a sender.

Of course, Mr. Tillerson is hardly the only person whose beliefs
(or at least those he purports to believe) are suspiciously aligned
with his motives. Other CEOs, like Steve Jobs, Elon Musk, and
Elizabeth Holmes, are famous for dispensing with reality. As one
of Jobs's coexecutives once put it, he suffered from "visionary optimism."[9] These executives are also famously persuasive; people
sometimes joke that they are surrounded by a "reality distortion
field." Some salespeople are similarly detached from reality in
their enthusiasm for their own company's products. If you've ever
had friends who've participated in multilevel marketing for Cutco
knives or Mary Kay cosmetics, you've probably noticed that they
use the products themselves—a decent sign they've internalized the
companies' spin—and often continue to do so for many years after
they stop participating in the program. Yet, no serious connoisseur
of knives or cosmetics promotes these products.

The role of persuasion motives in shaping beliefs becomes even
more stark when we look at people *within* a company, who just face
different motives. If you've ever interacted with a lawyer in your
organization's Office of General Counsel (OGC), or some other
individual who is responsible for assessing risk, you've probably
noticed that these individuals—whose job it is to reduce the risk
that their fellow employees expose the company to—are far more
risk averse than you and most others in the organization. There is
even the old joke that OGC is where proposals are sent to die.

OGC is hardly the only member of an organization whose risk
preference reflects its motives. Readers who remember the explosion of British Petroleum (BP)'s *Deepwater Horizon* oil rig in the
Gulf of Mexico might recall how, in the subsequent investiga-
tions, it came out that management at BP and their contractor,

Transocean, cut corners, feeling certain tests and precautions were not necessary, most famously on the day of the explosion, when a visiting BP official directed the crew to replace the heavy mud usually used to stabilize the well's pressure with lighter seawater. The crew, who were the ones at risk from a safety accident—six of them ultimately died in the explosion—did not share management's carefree disposition. They resisted this particular shortcut fiercely and, in surveys commissioned by Transocean a few months before the blast, complained that management prioritized drilling over maintenance and safety. While the workers' concerns ultimately proved justified, each side's beliefs reflected their motives: Transocean management's beliefs reflected the cost to the company from stalling the project over safety concerns, whereas their workers' beliefs reflected the risk they preferred to avoid.

In the above examples, motives are not randomly assigned, and we cannot be sure the extent to which the beliefs are internalized. That is, in the examples we've given so far—and in most of the ones we're liable to find in real life—all we see is a correlation between people's motives and what they believe, or at least purport to. However, it's hard to tell the extent to which motives caused these beliefs. And it's hard to tell whether people actually internalized these beliefs. Fortunately, as the following experiments demonstrate, psychologists and experimental economists have clever tricks for showing both these things.

Anyone who has participated in a court trial has probably noticed that the lawyers prosecuting the case seem genuinely more convinced of the defendants' guilt than do the lawyers defending them. That's despite the fact that, if anything, the defense has access to more damning facts about their client than the prosecution does. Again, each side's internalized beliefs reflect their motives.

Linda Babcock, George Lowenstein, Samuel Issakaroff, and Colin Camerer set out to show that this relationship was causal— that the sides' divergent motives were what caused their beliefs to diverge—and also to ensure that the beliefs in question were indeed internalized.[10] The research team invited law students at the University of Chicago and the University of Texas at Austin to the lab and randomly assigned them to represent either the plaintiff or the defendant in a negotiation to settle a case about a motorcycle accident. Both sides were provided the same twenty-seven pages of testimony, police reports, and maps about the case, which was based on a real-world case. Before they were paired up and sent off to negotiate, the students were asked to report what they thought a fair settlement would be and to guess what the judge would award in the real-world case; they received a bonus if their guess was within $5,000 of what the judge actually awarded. They were also rewarded if they recalled various facts from the case. Their responses to these questions wouldn't be part of the negotiation, so they were a good measure of subjects' internalized beliefs.

Even though only a few minutes had passed since they'd been randomly assigned their persuasion motive, subjects displayed drastically biased beliefs. Relative to subjects on the defendant's side, those on the plaintiff's side said a fair settlement was, on average, $17,709 higher and guessed the judge would award $14,527 more. Each side also remembered more facts—on average, 1.5 more—that favored their side than those that favored the other side. These are exactly the biased beliefs that we'd expect to emerge if subjects were engaging in asymmetric updating and asymmetric search in response to their assigned role in the negotiation. Moreover, because of the way in which the beliefs were elicited (subjects were paid more if they got closer to the judge's guess and remembered more facts), we can be relatively confident that the beliefs in question were internalized.

The rapid internalization illustrated by Babcock et al. is also on display among high-school debaters, who often come to earnestly believe the side of the argument they've been randomly assigned to argue for. In another fantastic study, Peter Schwardmann, Egon Tripodi, and Joël van der Weele elicited debaters' beliefs about the case they had been assigned to debate in a variety of ways, for instance, by paying them to guess correctly about the truth or asking them if they would be willing to donate some of their earnings from the study to a charity aligned with each side's position—again, relatively reliable measures of people's internalized beliefs. The researchers elicited these measures at multiple points during the debate: before the debaters had been assigned a side, after they'd been assigned but before the debate, and after the debate. The researchers found that the debaters' beliefs started out indistinguishable but diverged as soon as debaters were assigned a side and stayed that way, even after debaters had had a chance to hear the opposing team's arguments during the debate.[11]

People's beliefs don't just covary with what they tend to be motivated to argue but also with what they are *currently* motivated to argue. Donald Trump illustrates this phenomenon in its extreme. In 2009, Trump signed on to a full-page advertisement in the *New York Times* urging legislation to combat climate change, but by the time he ran for president in 2016, his position on climate change had changed to "I don't believe it" or "I don't think there's a hoax. I do think there's probably a difference. But I don't know that it's man-made." What changed? In 2009, Trump was still enmeshed in New York's liberal elite.[12] In 2016, he was pandering to right-wing climate denialists. Fast-forward to 2021, when Trump simultaneously insisted that he had won the 2020 election and was the country's legitimate president and that he could not be impeached because he was "a private citizen." Which position he took depended on the point he was trying to make: *I won!* or *I'm not guilty!*

While Trump may be extreme in his tendency to believe his own spin, we all do the same thing. Think about what you tell yourself when you've just met a new love and contrast this with what you tell yourself about the same person right after a breakup. Sure, you've learned a few things about your partner in the intervening period. But it's also the case that at the start of the relationship, you had a strong incentive to minimize your partner's flaws when trying to sell him to your friends and family, whereas now you need to call on these very flaws when trying to displace the blame for the breakup and ensure your friends and family (and maybe the judge who assigns custody) take your side.

Let's now shift gears to the second strand of evidence suggesting that motivated beliefs are well explained by internalized persuasion: that our beliefs covary with how strong our motive to persuade is, relative to our motive to be accurate.

There are a few ways we can look into this. One is to ask what people believe when they don't just have a motive to persuade but also money on the line. While their former CEO's position on climate change may be biased, when it comes to assessing Exxon-Mobil's own projects—drilling in the Arctic, for instance—the company is surprisingly respectful of the science. In 1992, an internal report noted how melting polar ice could make exploration more affordable but would also cause the seas to be rougher and endanger some of the firm's infrastructure. We see a similar phenomenon among investors, normally a conservative bunch that might otherwise deny climate change but who decidedly don't: they use state-of-the-art climate models—all of which predict that the climate is changing—to price weather futures on the Chicago Mercantile Exchange. We even see this phenomenon emerge when it comes to vaccines. During the COVID-19 pandemic, antivaxxer beliefs ran wild but mostly among people below the age of sixty, who were at much lower risk from the disease.[13] Seniors had skin in the game, and their beliefs in vaccines reflected this.

Florian Zimmerman has provided some neat experimental evidence that shows how subjects can be made less biased when they are given a stronger incentive to be accurate.[14] In his study, subjects took an IQ test and were told how they ranked relative to other subjects. Then, subjects returned a month later and were asked to recall their rank. The experiment had three treatment groups. The control group was told nothing about why they would be returning in a month. When they returned, they tended to remember their rank if it was high but forget it if it was low, leading to rather biased beliefs. Meanwhile, subjects in the two treatment groups were provided a stronger incentive to be accurate. In one treatment group, they were told in advance that they'd be paid to accurately recall their rank. In the second, they were told nothing, but then, when they returned, paid much more than the control to accurately recall their rank. These are two different ways of increasing subjects' accuracy motive. Either way, subjects with low ranks were less forgetful in these treatments and, consequently, ended up being less biased.

Another way we can see how people's beliefs vary with the strength of their persuasion motive (relative to their accuracy motive) is to see what happens when we introduce a motive to persuade. For instance, Peter Schwardmann and Joel van der Weele had subjects perform an IQ test alongside three other subjects and asked them to guess whether their performance was in the top two among their group of four.[15] However, before guessing, half the subjects were told they would be paired with an employer and could earn fifteen euros if they could persuade the employer that they performed in the top two. Both groups were somewhat overconfident, but subjects in the employer group were 7 percent more likely than those in the control group to say they were in the top two. Other research teams have performed similar studies in which they randomly assign some subjects to persuade others, and consistently, they find that those who are paid exhibit more biased beliefs.

ALTERNATIVE EXPLANATIONS

Before we move on to the next chapter (on repeated games and altruism), we'd like to discuss two alternative hypotheses to the one we presented.

In her classic review paper, Ziva Kunda described motivated reasoning as "a tendency to find arguments in favor of conclusions we *want to believe* to be stronger than arguments for conclusions we do not want to believe."[16] (We added the emphasis.) People would, for instance, want it to be the case that they're smart or good looking. It feels good to think you're smart and good looking! So this is what people believe. This explanation, which we will refer to as hedonic hacking, is probably the most popular explanation for motivated reasoning among psychologists, economists, and political scientists. It is distinct from the explanation we have proposed: our story was about persuasion being internalized, not about fooling ourselves into feeling good.

Hedonic hacking is likely going on some of the time. We do get pleasure from believing things are going well and may sometimes be able to use our smarts to hack our hedonic system, just like we can use our smarts to take antidepressants or take photos of things we'll one day get pleasure from reminiscing about. We might, for instance, get a kick out of carefully planning out a vacation we'll never be able to afford to take. Or enjoy daydreaming about what it'd be like to be a gazillionaire—so much so that it motivates us to buy a lottery ticket. Why wouldn't we also be able to fool ourselves into imagining how great our romantic partners are or how good our job prospects will be? Seems reasonable.

But there are several reasons why we think the hedonic hacking story is not the main story when it comes to explaining our motivated beliefs. For one, hedonic hacking is liable to be overcome by learning and evolutionary pressures anytime mistaken beliefs lead to costly mistakes and don't come with commensurate benefits (in terms of primary rewards). Overconfidence, for instance, has a cost. It can lead us to be overly demanding and aggressive and to overestimate the salary we could get in another job or how easy it might be for us to attract another partner. This can lead us to lose friends, jobs, and romantic partners. For people to not correct overconfidence, it must have some real, nonhedonic benefits, similar to the benefits from persuasion that we have been discussing.

Another problem with the hedonic hacking story is that we don't always fool ourselves into believing what we want to be true. For instance, if you randomly assign subjects to one side of an argument or the other, and they both come to believe the side they were assigned, then it can't be that they all are believing what they want to be true. Moreover, there are plenty of things we would want to be true that we don't fool ourselves about. Michael Thaler has shown that we don't tend to be overconfident about things that we might want to be true but that don't help us persuade others of anything desirable about ourselves.[17] He asked subjects about good and bad things facing humanity as a whole, like cancer survival rates. We *want* it to be true that these numbers are low but don't have any particular reason to convince others they are. He found people are not overly optimistic about these kinds of things, further pointing to persuasion as the key driver of biased beliefs.

Sometimes, we even fool ourselves into believing things we decidedly wouldn't want to be true but that are beneficial when it comes to persuasion. Consider those who are assigned to argue for nuclear weapon bans, who then make the case that unlimited nuclear arsenals will inevitably lead to armageddon. Do these people really want it to be true that we are on the path to annihilation? It's likewise not the case that the lawyers at OGC *want* risky things to come true. When American conservatives overestimate the increase in the crime rate under President Obama, is this because they want there to be more crime?

Such examples are, in our opinion, better explained as people believing what they want to persuade others of: those arguing in favor of nuclear limitations want to persuade others of the importance of limiting nuclear arsenals, OGC wants to persuade others to reduce risk, and conservatives want to persuade others to reduce immigration or that Democrats don't make for good presidents. Heck, we're not even always overconfident. When we're trying to convince others we're not a threat or need their help, we're suddenly far more likely to be full of doubt or even self-pity. The fact that we can be made to be underconfident is also better explained by the persuasion story than the hedonic hacking story, we think.

A third problem with the hedonic hacking story is that it leaves unexplained the *ways* in which we fool ourselves. Motivated reasoning involves some glaring asymmetries, such as paying attention to the results of a search but not the extent of it. Why would it be so easy to ignore the latter but not the former? The persuasion story explains such asymmetries on the basis of what *other* people tend to know or are liable to find out and sanction us for. But if you are just trying to fool *yourself*, then why would you be aware of what you found but not of how hard you searched?

Combined, these three problems with the hedonic hacking story lead us to believe that while it may contribute to biased beliefs now and again, it's not the primary driver of motivated reasoning.

Aside from the hedonic hacking story, the most common alternative explanations that we encounter for motivated reasoning have nothing to do with people's motives. They are things like people are lazy[18] or not such great Bayesians even when they try. Or they're actually pretty good Bayesians, but the information they have access to and trust is really different. Their Facebook feeds are echo chambers. Everyone around them only listens to Fox News and keeps insisting that CNN is a bunch of baloney.

Like the hedonic hacking story, these nonmotivational accounts are intuitively appealing and likely do explain some of the biased beliefs we encounter but, we think, can't be the whole story. Internalized persuasion has to be part of the picture.

Here is the biggest limitation of the nonmotivational accounts, in our opinion. Even though people have access to (or trust) different sources, you'd expect them to eventually learn that their beliefs are off-kilter and correct them. Why don't they update their beliefs when they encounter people who disagree with them or have access to different sources? The fact that there is disagreement is informative, so why don't they respond to this information? This argument was introduced by Nobel Laureate Robert Aumann in

"Agreeing to Disagree" and expanded on by John Geanakplos and Heraklis Polemarchakis in "We Can't Disagree Forever." To us, it is the Achilles' heel of nonmotivated accounts. At some point you need *some* motivational account to account for the fact that there is no updating.

That said, another limitation of the nonmotivational accounts is that they, again, don't address the asymmetrical ways in which people update beliefs: that they update when they get supportive evidence but not when they fail to get unsupportive evidence, or that they update when they find supportive evidence but not for how they've searched. Why are people just fine at incorporating supportive evidence but awful at incorporating lack of unsupportive evidence or how they searched for evidence? A lazy person should be equally likely to ignore supportive evidence as he is ignoring how it was obtained. A Bayesian should be just as likely to update her beliefs downward as she is to update them upward. And if she searches harder or cherry-picks, then she should devalue what she finds.

When it comes to designing experiments that might distinguish these nonmotivational accounts from internalized persuasion, things can get thorny pretty fast. The issue is that any given source that goes against someone's priors ought to be suspect. One study, also by Michael Thaler, attempts to get around this concern.[19] Thaler began by asking subjects questions like: How much did the crime rate increase under Obama? How likely are Black people to be called back for interviews relative to white people? The statistic we mentioned earlier—that conservatives are more likely to believe crime went up under Obama than liberals—is from this study. Thaler similarly found that liberals are more likely than conservatives to believe that Black people are less likely than white people to be called back for interviews.

Then, Thaler demonstrated that the way subjects update their beliefs is asymmetric and can't be explained by how much they distrust a source. Here's how he did this. After participants responded to his questions, Thaler showed them a message that told them that the correct answer was either above or below their guesses. However, there was a catch: with 50 percent likelihood the message was telling the truth, but with 50 percent likelihood it was a lie. Thaler told them this and then asked them to guess whether the message was true or false. He paid them a bonus if they guessed correctly. A good Bayesian would say it's fifty-fifty. After all, she had given her best guess already and should expect the truth to be equally likely to be above this guess or below it. And yet, participants were more likely to say the message was false if it corrected their beliefs in a way that was inconvenient for their politics: conservatives were more likely to say the message was false if it said that their response about crime rates under Obama was too high, and liberals were more likely to say the message was false if it said their responses about call-back rates for Black candidates was too low. Thanks to Thaler's clever design, we can be confident that this bias is due to participants' persuasion motives and not their priors about sources' reliability. The nonmotivational accounts have a tough time explaining this result.

So that's our case: motivated beliefs are about internalized persuasion, at least to a large extent. The aspects that can be explained by nonmotivational accounts or hedonic hacks might explain some of our biased beliefs, but the internalized persuasion story is still essential.

CHAPTER 10

THE REPEATED PRISONER'S DILEMMA AND ALTRUISM

THE (ONE-SHOT) PRISONER'S DILEMMA IS THE MOST FAMOUS GAME IN town. In this game, two players simultaneously choose whether to cooperate or defect. Cooperation is costly, but benefits the other player by more than it costs; it costs c, and benefits the other $b > c$. Although it would be better for everyone if both players cooperated, defecting is a dominant strategy—each player is always better off defecting, regardless of what the other is doing. This makes mutual defection the unique equilibrium.

The prisoner's dilemma is the simplest way of illustrating the distinction between individual self-interest and the collective good and showing that the two don't necessarily correspond. In the game, the unique Nash equilibrium is not the socially optimal solution. Both players could have gotten $b - c$, but instead, they got nothing.

And yet, people do often cooperate. They're good to each other. They help one another. What gives?

A common response is to say game theory is wrong. People are not rational. They don't have selfish preferences. They cooperate because they care about each other. Hopefully, by this point in the book you will agree with us that this isn't the most satisfying resolution. Yes, proximately, people care. But ultimately, this caring is

169

coming from somewhere. Can we use game theory to better understand where it's coming from and how it works?

The answer lies with modeling repeated interactions. In the one-shot prisoner's dilemma, players interact once and then it's game over. There is no way for their cooperation to return dividends in any way. That's not true in reality. People don't typically interact just once, in a vacuum. There's typically some chance they'll meet each other again unexpectedly, that the interaction will be observed, or that word will get out. Can this help motivate cooperative behavior (and the corresponding altruistic sentiments and moral intuitions)? If so, how? What would this require?

To answer such questions, we need a model. Two, actually. We'll start with the repeated prisoner's dilemma—the classic model of dyadic interactions (interactions between the same two players). Then, in the next chapter, we will expand our analysis so that, instead of just two individuals repeatedly interacting, there is a community of individuals.

THE REPEATED PRISONER'S DILEMMA

Here's how the repeated prisoner's dilemma (RPD) works. In each round, starting with the first round, players play the plain old prisoner's dilemma. They choose whether to cooperate and get their payoffs. Then, they learn what the other player did and play again. Each new round gets discounted by δ. δ is traditionally thought of as reflecting the fact that payoffs (like money) tomorrow are less valuable than payoffs today. But it can equally as well be thought of as capturing the uncertainty over whether the game will repeat.

The point of the RPD is to characterize ongoing dyadic relationships (relationships between a pair or two cohesive groups) in which those involved occasionally can benefit from a bit of help from their partner. Maybe they're a pair of hunters, and sometimes one comes up dry when the other has had a string of luck. Maybe they're a pair of students—one good at math and the other at history, so each can help the other understand some material better.

Maybe they're a pair of neurotic coauthors who can occasionally use each other's help evaluating a new idea, editing an article, analyzing some data, or drafting a chapter of a book. If their interaction is expected to be long lived—δ is closer to 1—or the mutual gains they get from cooperating with each other is high enough—the ratio of b to c—it will be possible to sustain cooperation. But to see that, we first need to describe how such games are analyzed.

A strategy in a repeated game (like the repeated prisoner's dilemma) must specify what a player does in any given round—cooperates (plays C) or defects (D)—given *everything* that's happened in previous rounds. Here are some possibilities for this game: Players can defect in every round, regardless of what's happened in the past (this strategy is sometimes called ALLD). Players can also cooperate in every round, no matter what's happened in the past (ALLC). They can also cooperate only in some predetermined rounds (say, the even ones). Or they can play a conditional strategy, in which they respond to what's happened so far. Here are two well-known conditional strategies that we'll be discussing in some detail:

- Grim trigger: Cooperate in the first round, then continue to cooperate so long as neither you nor your partner has ever defected. If anyone ever has defected, pull the trigger and play D from that point on.
- Tit-for-tat (TFT): Cooperate in the first round and then, from that point forward, simply copy what your partner did in the previous round.

EQUILIBRIA OF THE REPEATED PRISONER'S DILEMMA

There is an easy way to check if a pair of strategies is a Nash equilibrium. We just need to check if either player can benefit by deviating once, then following their prescribed strategy forever after. If no "one-shot" deviation can benefit them, more complicated deviations won't either.

Let's use this one-shot trick to look into whether the strategies we already discussed—ALLD, ALLC, TFT, and grim trigger—are Nash equilibria, starting with ALLD.

If the other player is always defecting, can you ever benefit by cooperating? No, you'd simply be paying c, and the other player would never care; he is defecting regardless. Playing D is the best you can do. Since the same is true for the other player, voilà, ALLD is an equilibrium.

Now, say you and player 2 are both playing the strategy ALLC. Well, what if, in some round, you deviate from C to D? You save c in that round. Meanwhile, nothing changes down the road: your opponent is playing ALLC, so is going to cooperate regardless of what you did. We have found a beneficial one-shot deviation. This ain't Nash.

Next, let's try grim trigger. First, let's check what happens when you both play grim without deviating: in the first round, you both play C. In the second round, no one has defected, so you both play C. Same in the third round, and so on. What changes if you were to pick an arbitrary round and deviate to D in that round? You'll save c in that round. But, then, both you and your partner pull the trigger and play D forever after, meaning you'll forego the benefits of cooperation forever more, which add up to $(b - c)(\delta + \delta^2 + \delta^3 + \ldots)$. Would you benefit from this deviation? Not if $c \leq (b - c)(\delta + \delta^2 + \delta^3 + \ldots)$. A bit of algebra, and this condition simplifies to $\delta \geq c / b$. And that's our key equilibrium condition. Cooperation can be sustained if the likelihood of a repeated interaction, δ, is sufficiently large in comparison with the relative costs of cooperation (relative, that is, to its benefits), c / b.

What about TFT? If you both stick to the strategy, then, in the first round, you both cooperate. In the second round, each of you cooperates because that's what the other did in the last round. Same with the third round. And so on. Again, we're seeing co-operation all the time. What happens if you choose an arbitrary round for a one-shot deviation to D, then continue playing TFT ever after? In the round that you deviate, you will gain c. In the next round, you will cooperate and the other will defect—playing tit for your tat. After that she will cooperate, and you will defect. Then, you will keep oscillating in this way for eternity. Again, we see the same trade-off as we did for grim: defect now, and you save now, but it costs you down the road. Indeed, if you do the math out, you'll find that, once again, this deviation makes players worse off so long as $\delta \geq c / b$.

SUBGAME PERFECT EQUILIBRIA

Game theorists often talk about subgame perfection when they deal with repeated games like the prisoner's dilemma, so we'd like to take a moment to introduce the concept.

Subgame perfect is like Nash but a little stricter. It requires that players not only can't benefit by deviating from their prescribed strategies, but they also can't benefit by deviating from their pre-scribed strategies after sequences of events that aren't expected to happen. For instance, if both of you play TFT, neither of you ever expects the other to defect. And, as we saw, when $\delta \geq c / b$, neither of you can benefit from deviating. But would you be better off deviating from TFT if someone did defect? Yes. You would be better off ignoring that defection (again, provided $\delta \geq c / b$) because then you can go back to both cooperating in each round, instead of constantly oscillating between cooperation and defection. TFT is Nash but not subgame perfect.

Grim trigger and ALLD, on the other hand, are both subgame perfect. In grim, we check this by looking at how you'd respond to a defection (which isn't supposed to occur). Would you be better off ignoring it? No, since once it occurs, your partner will play D forever more no matter what you do. In ALLD, we check how you'd respond to cooperation (again, not supposed to happen). Should you switch to cooperating? No use, since your partner will soon enough go back to defecting forever more, regardless of what you do.

Remember that we justified the use of the Nash equilibrium by arguing that evolutionary and learning processes tend to lead to

Nash. The logic was just that when you aren't at Nash, someone would benefit from deviating (and so he or his followers or descendants would learn or evolve to do so). But if a strategy profile is Nash, there would be no evolutionary or learning pressures to deviate.

A similar argument can be made for subgame perfection. We just need to recognize that the world is noisy, so that every once in a while players find themselves in histories that weren't expected to occur (for instance, if a player who evolved to play tit-for-tat accidentally plays D even though her partner did not defect in the last round). So long as players face some pressure to behave optimally even in these histories, then they'll arrive at an equilibrium that's not only Nash but also subgame perfect.[1]

All the insights we are going to cover in this chapter rest on Nash, not on subgame perfection, so for now, we're not going to harp on the distinction. But you'll soon see subgame perfection adds a lot of insight. We'll use it heavily in the next chapter and again in Chapter 14, which covers some features of our sense of justice.

THE NECESSARY FEATURES OF COOPERATION

Although we have analyzed just a tiny fraction of the strategies and equilibria in the RPD (there are uncountably many), we've already seen enough to identify some recurring features in all strategy profiles that can sustain cooperation. These features are key to understanding how altruism actually works and will also be apparent in the model we discuss in the next chapter. They're also hard to explain with more proximate or emic approaches.

δ **must be high.** We saw this in the two equilibria we analyzed that sustained cooperation (grim and TFT). We can also see this by the comparing to the extreme case of $\delta = 0$, the one-shot prisoner's dilemma, where mutual defection is the only equilibrium. Finally, we can see this by noting that anytime you cooperate you pay a cost today, c. For you to not benefit by deviating, there has to be some benefit down the line, but that can only matter to you if δ is sufficiently high.

Reciprocity. Think back to our analysis of ALLC, which was not an equilibrium. The fact that both players cooperate unconditionally is (proximately) heartwarming, but (ultimately) there is no incentive for them to keep doing so. That incentive comes in the form of a threat that cooperation will be withdrawn when the other player defects or will only be forthcoming if the other player cooperates. This is something we see with both TFT and grim. This feature is often called reciprocity or reciprocal altruism. It is also sometimes referred to as conditional cooperation.

Expectations matter. As we've seen, the RPD has many equilibria. Some involve cooperation (like grim and TFT) and some don't (like ALLD). Some involve more forgiveness than others (grim is pretty unforgiving, TFT ignores whatever you did more than one round back). How do players know what to do? They have to have some way of assessing which equibrium others are playing. This means that, much as we saw in the hawk-dove game, people's behavior can be drastically affected by anything that shifts their shared expectations.

Higher-order beliefs matter. Here's one final feature of cooperative equilibria. Imagine you and your partner are playing the RPD, and you're both playing some strategy that supports cooperation in equilibrium—say, grim trigger. Suppose that, a few rounds in, you notice your partner means to cooperate, but without realizing it, he defects. Do you play as prescribed by grim and pull the trigger, defecting henceforth? Nope. You are better off pretending nothing happened. Thus, it's not just your first-order beliefs—whether you think the other player defected—that matter. Your second-order beliefs—whether you think the other thinks he defected—matter as well. The same can be said about higher-order beliefs: if you think the other noticed that he defected, but you also think that he thinks you think he didn't notice, you have an incentive to overlook the defection then as well.

SOME EVIDENCE

For the rest of the chapter, we will focus on providing a bit of the evidence for the first two features listed above: high δ and reciprocity. We're going to save the discussion of evidence that expectations and higher-order beliefs matter for future chapters since a lot of the most interesting evidence isn't particular to dyadic interactions.

Let's start by discussing the evidence that, for there to be cooperation, δ must be high. In *The Evolution of Cooperation*, Robert Axelrod provided a (now-famous) example of how altruism can arise when δ is high, even in the most unforgiving of environments: the trenches of World War I. Because of the technologies available to the armies of the time, trenches were an effective means of stopping an army's advance; the best way over a trench was around it. So, at the start of the war, both sides frantically dug longer and longer trenches in what became known as The Race to the Sea, until a network of trenches extended from the North Atlantic in Belgium to the Alps in the south. Then things ground to a halt, and the armies settled in for what promised to be a long and brutal showdown.

Paradoxically, though, the standoff meant that the same units faced off against each other for months at a time. This, in turn, meant δ was high and opened the door for cooperation between the two opposing sides. Sure enough, in many places, both sides stopped fighting for real and instead feigned sniping and bombardments by intentionally shooting at the same places over and over. This allowed the soldiers to avoid being reprimanded by their superiors, while simultaneously making it completely straightforward for their "enemies" to avoid getting harmed and to benefit from the same treatment. There are even records of soldiers apologizing—at great personal risk—for accidentally firing a round at a place and time that wasn't expected.

The cooperative bout only ended when the commanders on both sides finally wised up and began rotating the units so that no unit stayed in the same place too long. This reduced δ and led to cooperation breaking down and the fighting returning in earnest.

Next let's check whether, in practice, altruism is conditional— whether we are more likely to cooperate with those who have co-operated with us in the past.

One of the most famous examples of conditional altruism comes not from humans but from vampire bats. Vampire bats are prone to starve if they fail to hunt successfully for a few nights in a row, but they have a tendency to help one another. When a bat returns from a successful hunt, it will sometimes regurgitate some of its blood into the mouth of a bat whose hunt was unsuccessful. In a series of studies in which he randomly selected bats and starved them, Gerald Wilkinson showed that a previously starved bat eagerly repaid the favor to a bat that had helped him when that bat was starving.[2] (He also took pains to provide evidence that δ was sufficiently high amongst vampire bats to support cooperation by documenting that their broods were very stable. Some bats had been members of the same brood for nearly a decade.)

Robert Axelrod argued quite forcefully that, as heartwarming as the cooperation between the two sides in the World War I trenches was, it was also highly conditional. Soldiers reported that in the event that the other side did shoot or bombard in earnest, they'd respond by hitting them back three times as hard. They also made it very clear that their punishment would have teeth. Snipers would shoot at the same spot in a wall with such accuracy that they would bore holes in the walls, sending a clear signal of what would happen if they aimed at a soldier instead of the wall.

WE'VE GOT SOME MORE FEATURES TO DIG INTO, BUT BEFORE WE DO, let's extend our analysis beyond dyadic relationships.

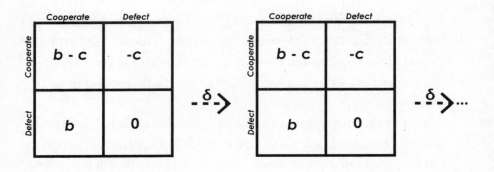

SETUP

- There are two players. In each round, a prisoner's dilemma is played. With probability δ they continue on to the next period. Otherwise the game ends.
- In the prisoner's dilemma, the players choose between two options: cooperate (C) and defect (D). If they cooperate, they pay a cost $c > 0$, and the other benefits by $b > c$. If they defect, they pay nothing, and the other gets nothing.
- In this model, players can observe everything that's happened in the past.

SOME STRATEGY PROFILES OF INTEREST:

- Tit for tat: in round 1, play C. In every subsequent round play whatever the other player played in the last round.
- Grim trigger: in round 1, play C. In every subsequent round, play C except if anyone has ever defected, then play D.
- Always defect: play D in each round no matter what has happened in the past.

EQUILIBRIA CONDITIONS:

- Cooperative equilibria like tit for tat and grim trigger can only be sustained if $\delta \geq c / b$.
- Always defect can always be sustained.

INTERPRETATION:

- To sustain cooperation:
  - There must be a high likelihood of "repeated interaction" and past behavior must be highly "observable" (both are represented in the model by δ).
  - Cooperating today must yield some future benefit, like others being more likely to cooperate with you later (reciprocity). Thus, cooperative behavior needs to be conditional: people can't just cooperate indiscriminately, as they might with kin or if they were directly motivated to improve others' welfare.
- Because there are multiple equilibria and, in particular, there is always an equilibrium in which players always defect, people will be sensitive to expectations, context, framing, and so on—anything that might influence whether they expect cooperation to be rewarded.

CHAPTER 11

NORM ENFORCEMENT

THE REPEATED PRISONER'S DILEMMA LEAVES AN IMPORTANT QUES-tion unanswered. The game has just two players and is thus well suited for understanding cooperation between two individuals or groups—friends, coworkers, or even two opposing World War I army divisions. How can we use it to understand what's needed to get people to do to things that benefit society more broadly, for instance via charitable donations, volunteering, resource conservation, and so on?

To answer this question, we will introduce a simple model of norm enforcement.[1]

In the model, there are n players, where n is at least 2 and as big as you'd like it to be. In the first round of the game, each player is randomly chosen to make a choice: comply or shirk. Complying carries some personal cost C. It will often be interpreted as bene-fiting the group but need not.[2] It represents things like contributing to the church offering, donating to a museum, volunteering at the church soup kitchen, reusing your towel, or helping an old lady cross the street, as well as discriminating against outgroup members or not driving on shabbat.

With probability δ, we move on to the second round. In the sec-ond round, all the players are randomly paired, and each chooses whether to punish the player they are paired with, which involves paying a cost c to harm that player by h. Then, with probability δ, we move on to the next round, in which players are randomly paired and can pay a cost to punish each other, once again. And so on. For simplicity, we assume that everyone knows everyone's past behavior.

Just as ALLD was a (subgame perfect) equilibrium of the repeated prisoner's dilemma, this game has an equilibrium in which players don't comply in the first round and never punish in subsequent rounds. After all, if no one will punish you for failing to comply, why would you? And since punishment is costly, if no one will punish you for failing to punish, why would you?

Now consider this alternate strategy profile, which involves complying:

- In the first round, comply.
- In the second and all subsequent rounds, punish only those who didn't comply in the first round or didn't punish someone who should have been punished in the previous round.

This strategy is a (subgame perfect) equilibrium so long as δ is sufficiently high—technically, so long as it is greater than both c/h and C/h. How do we know? As usual, we check if anyone can benefit from deviating, assuming no one else does (now making sure to also check cases where someone finds themselves in unexpected circumstances). So, would you want to shirk in the first round? If you do, you'll gain C, but then, in the next round, others will punish you, harming you by h. As long as δ is big enough, this harm will more than offset the gains you got from shirking, even after being discounted. OK, but what if you found yourself (unexpectedly) paired with someone who had shirked or had failed to punish when she was expected to? You're supposed to punish her. Should you? Well, you could save c by not punishing, but this means you'll get punished in turn. That'll cost you h. Again, as long as δ is big enough, this harm will more than offset the savings, even after being discounted, leaving you worse off. Voilà, we have ourselves a subgame perfect Nash equilibrium.

◆

THIS EQUILIBRIUM OF THE NORM ENFORCEMENT GAME SUGGESTS two features of norms we can look out for.

Third-party punishment. The first is third-party punishment. Compliance in the first round of the norm enforcement game is motivated by the threat of punishment in future rounds. The players doing the punishing are third parties, so-called because they punish even though they aren't directly harmed if the first player fails to abide by the norm.

Higher-order punishment. The second feature is higher-order punishment. To ensure an equilibrium strategy in the norm enforcement game is subgame perfect, we not only need to incentivize compliance but also punishment. This is done via higher-order punishment—punishment of those who fail to punish when they should have.

In addition to these two features of norms, some of the features we saw in the last chapter still play a significant role.

Observability. If people can't observe your misdeeds, they certainly can't punish you for them. This may occur if behavior is sufficiently anonymous or social networks are not tight enough to motivate continued monitoring and enforcement. Broadly speaking, one can think of all these factors as related to δ; that is, a high δ doesn't mean just that the likelihood of interacting again is high but also that those we interact with will know what happened in the past and themselves have sufficient incentive to care.

Expectations. Like the repeated prisoner's dilemma, the norm enforcement game has multiple equilibria: some with compliance and some without, and also some where one norm is enforced and some where another is enforced. So, you'll want to assess which norms your community expects you to abide by and enforce— and when.

Higher-order beliefs. Just as a player would prefer not to punish an unwitting defection in the repeated prisoner's dilemma, so a player in the norm enforcement game would prefer not to punish a norm violation that only she witnessed, as she then needlessly bears the cost of punishing, and risks being punished for it.

◆

LET'S TAKE A LOOK AT SOME EVIDENCE. WE'LL START BY DISCUSSING evidence for the two features on this list most unique to norm enforcement—third-party and higher-order punishment—and then discuss evidence for the other items on this list.

Third-party punishment. Do people actually third-party punish norm violators whose violations don't directly hurt them? Here are a handful of examples that suggest they very much do.

For our first example, we'd like to ask you to conduct an experiment. For the sake of science. The next time you're in a public place, take a candy wrapper and throw it on the ground in front of someone. Actually, don't do this—Moshe already did it for you when he was an impish teenager. The result? A burly third party stopped his pickup truck in the middle of the crosswalk, got out, and let him know that it would be in his interest to pick the wrapper up. It's worth emphasizing that the burly third party was not directly impacted by Moshe's litter; it was in a public street, and the third party just happened to be driving by. (Moshe learned his lesson and has since internalized the antilittering norm.)

For our second example, we'll look at some real experiments, which were designed by Ernst Fehr and Urs Fischbacher.[3] In these experiments, one subject plays an ultimatum game with a second subject. Then, a third subject is asked whether he'd pay to reduce the first player's payout. Fehr and Fischbacher found that the answer was yes: if the first subject failed to give generously in the ultimatum game, the third subject, who was in no way affected by the first subject's selfishness, was nonetheless willing to pay to punish her. Similar experiments show that subjects are willing to punish those who fail to contribute to a public good—like paying into a pot that the experimenter will double and divide up among everyone in the group.[4]

A large team of researchers led by Joe Henrich used Fehr and Fischbacher's experiment to explore just how ubiquitous third-party punishment is.[5] They recruited subjects from diverse cultures— nomadic and seminomadic tribes in East Africa and the Amazon, farmers in Papua New Guinea, Tahiti, and Missouri, and students at Emory. In all these places, they had their subjects play Fehr and Fishbacher's third-party punishment game, with an endowment of a whole day's wages.

Henrich and his team found that there was quite a bit of variation in the degree to which the first subject was generous to the second subject and the third subject was willing to engage in third-party punishment. However, the variation wasn't random: the places where the first subject was more generous were also the same places where subjects engaged in third-party punishment. Moreover, even in the least cooperative locales, they found that a meaningful share of subjects third-party punished (and, consequently, a meaningful share of subjects were willing to share their endowment).

Our third example of third-party punishment comes to us from the Turkana, a tribe of nomadic pastoralists in East Africa. Through careful ethnographies, Sarah Mathew and Rob Boyd document the use of third-party punishment to encourage participation in cattle raids among the Turkana. Desertion during a raid is reasonably common but not as common as you might think, given that the raids account for a fifth of all male deaths. That's because desertions are punished via public tongue lashings, fines, and, in extreme cases, public lashings in which a deserter is tied to a tree and beaten by his peers. Luke Glowacki found similar customs among another East African tribe, the Nyangatom.[6]

Our fourth and final example comes to us from the Jim Crow South. The historian Clive Webb writes that, during this period, immigrant merchants in the South were often not especially biased toward Black people and benefited from their business but nonetheless towed the racist line. Webb quotes a shopkeeper in 1920 as saying, "I'm here for a living, not a Crusade," and says that those who weren't as careful faced "economic reprisals, social opprobrium, and even violence. On August 15, 1868, a young Russian Jew named S. A. Bierfield was seized by Klansmen and shot dead. Bierfield had caused offense to the white folk of Franklin, Tennessee, by fraternizing with the Blacks who shopped at his store. Fourteen months later, Samuel Fleishman, a Jewish hardware merchant, was murdered in almost identical circumstances in Marianna, Florida."[7] Crucially, the white folk committing these violent acts were third parties who weren't directly affected by the shopkeepers' extra sales or fraternizing.

Higher-order punishment. Does higher-order punishment actually motivate third-party punishment? Often, yes. Here's a smattering of evidence.

Rob Kurzban, Peter DeScioli, and Erin O'Brien have shown that people are more likely to third-party punish when they know they're being watched, which is a simple way of triggering the feeling that one might be higher-order punished.[8] They brought subjects into the lab for a two-part experiment. In the first part, two subjects were asked to play a variant of the prisoner's dilemma, in which the players choose whether to cooperate sequentially—first subject one decides whether to cooperate or defect and then subject two decides the same. As usual, subjects earned the highest payoff by defecting against a partner who cooperated. In this case, they could make thirty dollars! In the second part of the experiment, a new, third subject was brought in and told how the players had played in part one. Then, she was given an endowment of ten dollars and asked how much she would like to pay to punish subject two. For every dollar she put toward punishing, this reduced subject two's payment by three dollars. The expectation was, of course, that subject three would (only) punish subject two if he had defected against subject one, who had cooperated.

Kurzban, DeScioli, and O'Brien assigned subjects in the second part to three treatment groups. The first completed the experiment completely anonymously. Even the experimenters couldn't link their decisions to their identities. The second treatment group's decisions were known only to the experimenters but kept secret from other participants. In the third treatment group, subjects were told they'd have to stand up and announce their decision to the remaining subjects at the end of the experiment. Sure enough, the less anonymity, the more third-party punishment subjects exhibited. When the experiment was fully anonymous, average punishment amounts were $1.06. When the experimenter could learn their decisions, subjects punished $2.54, on average. And when other participants learned their decisions, subjects punished $3.17, on average. When a participant knew that others would know whether he had punished and could, at least in theory, higher-order punish him, he was motivated to engage in substantially more third-party punishment.

In a more recent study, Jillian Jordan and Nour Kteily repeated Kurzban et al.'s experiment but with a twist.[9] They replaced the first part of the experiment with a vignette, in which they

> gave subjects a chance to support a group of organizers, who were attempting to punish a university professor following sexual harassment allegations against him. Subjects learned that the organizers were "marching with megaphones outside of [the professor's] office and personal residence in order to expose his behavior to his friends, family, neighbors, and professional network, and mount pressure on [the university] to take action against him." Subjects then decided whether to punish the professor by giving money to the organizers, and also reported (on continuous scales) their commitment to supporting the organizers and their personal moral evaluations of the case for punishment.

As Kurzban et al. had done, Jordan and Kteily varied whether subjects' punishment decisions would be observed. The key twist is that they also varied just how sure subjects would be that the professor had indeed done wrong. Here's how they describe it:

> We were able to manipulate the ambiguity of the moral case for punishment. We presented subjects with a news article describing the allegations, and varied important details surrounding their credulity and severity. In the unambiguous condition, the allegations were very severe, and very likely to be true (e.g., the most severe accusation was of attempted rape; the professor did not deny the accusations; there were six accusers). In the ambiguous condition, the allegations were relatively less severe and less likely to be true (e.g., the most severe accusation was of relatively more minor unwanted touching; the professor denied the accusations; there were two accusers, both of whom had potential ulterior motives; sources vouched for the professor's character).
>
> Despite these important differences, subjects in both conditions learned that the organizers were employing the same (relatively severe) punitive approach.

In the unambiguous case, Jordan and Kteily's findings were much like Kurzban et al.'s: subjects were meaningfully more likely to punish when they were observed.

Jordan and Kteily also found that, as expected, subjects in the experiment punished substantially less often—about a fourth as often—in the ambiguous case. But they nonetheless found that subjects punished more when they were observed: "in a striking demonstration of reputation's motivational force, subjects in the ambiguous condition were no less responsive to our reputation manipulation." The implied threat of higher-order punishment was so great that it caused subjects to overcome their own reservations and go ahead and punish. Striking indeed.

So people are indeed motivated to third-party punish by the threat of higher-order punishment. But do people higher-order punish? Do they really go after people simply for failing to punish when they should have? The answer to this question is also yes.

Sarah Mathew explored this question among the Turkana in surveys that described four different scenarios.[10] The first described someone who had punished a deserter. The second described someone who hadn't punished a deserter. The third described someone who had punished excessively, and the fourth described someone who had punished unjustly—that is, they had punished someone who hadn't deserted. The Turkana were then asked: "Was he wrong? Are you displeased with him? Was his behavior not useful? Would you criticize him? Would you punish him? Would you withhold help from him?" In the first scenario—the one in which the person in question had punished the deserter—the Turkana rarely replied that the individual in question was wrong, that they were displeased with him, and so on. However, for the remaining scenarios, they said so rather frequently, also indicating with some frequency that they would punish him. Punishing those who fail to punish deserters but not punishing those who punished deserters is exactly what the norm enforcement model requires.

When Jackie Robinson joined the Brooklyn Dodgers, breaking the so-called color line in Major League Baseball, it raised the ire of some of the league's more racist fans, most famously, those of the St. Louis Cardinals. Robinson, of course, bore the brunt of their repeated abuse—both verbal and physical. His non-violent response would ultimately serve as an inspiration to the Civil Rights Movement. Robinson's teammates were also targeted, despite the fact that they had nothing to do with the decision to hire Robinson and some had even opposed it. But, by staying on the Dodgers, they had, in the estimation of racist fans, failed to punish when they should have and thus had earned their higher-order punishment.[11]

Both Christians and Jews practice excommunication (in Judaism, it is called *herem*). If a community member misbehaves, he is excommunicated by the religious leadership, which means that not only is he prohibited from participating in the community but also that anyone who associates with him can be excommunicated. The Fourth Lateran Council of the thirteenth century, for instance, gave clergy the power to excommunicate members who hadn't themselves committed heresy but had come to the aid of someone who had.[12] As a middle schooler in Israel, Erez witnessed schoolchildren practice a nonparochial form of *herem*. When a classmate committed a particularly severe faux pas, his or her peers would declare a *herem*. Then, anyone else who violated the initial *herem* by so much as talking to the banished classmate would themselves be subjected to a *herem*. Amazingly, the kids involved were mostly secular and had likely never heard of the parochial adult *herem* after which their own *herems* were named.

The psychologists Kylie Hamlin, Karen Wynn, Paul Bloom, and Neha Mahajan have documented that even infants engage in what looks like higher-order punishment.[13] The team used sock puppets—elephants, a moose, a duck—to act out two scenes. In the first, one puppet was shown being either nice (it helped another puppet open a box containing a toy) or mean (it slammed the lid of the box shut). Then, that puppet was, itself, shown in need (it was playing with a ball and dropped it). Another puppet came on the scene and either gave the ball back or took it away. Finally, the infants were allowed to choose a puppet to play with. At five months, infants universally chose the puppet that gave the ball back to the first puppet, regardless of whether that puppet had been mean or nice in the first scene. But by eight months, infants second-order punished. They chose the puppet who was nice to the puppet who had been nice, and the puppet who was mean to the one who had been mean.

NEXT, LET'S TAKE A LOOK AT SOME EVIDENCE FOR THE THREE FEATURES of cooperation in the norm enforcement game that were also shared with cooperation in the repeated prisoner's dilemma.

OBSERVABILITY

The first feature we'll check for is observability—that people are more altruistic when observers are more likely to find out about their good deeds and when they care more about those observers.

The first evidence we'll consider comes from a simple laboratory experiment that's often used to study altruism known as the dictator game. In the dictator game, subjects are given an endowment of a few bucks, then asked what share of this endowment they're willing to give to another subject. And that's that. That's the whole experiment. In this simple experiment, many subjects give half, though some give less and others give nothing at all.

Consistent with our prediction, how much subjects give depends on how anonymous the experiment is. If subjects' identities are revealed to other subjects, subjects tend to give at very high rates. If their identities are withheld from other subjects, they become less generous. If the experiment is double blind—meaning their identities are withheld not only from other subjects but also from the experimenter—they become even less generous. If the experimenter employs extraordinary measures to further protect the identities of subjects and emphasizes these measures—really hitting subjects over the head with just how anonymous the experiment is—then, finally, they give almost nothing. The bottom line is, the more observable things are, the more subjects give.[14]

In addition to the dictator game, another common laboratory paradigm for studying altruism is the public good game. This paradigm is a bit like a dictator game but with more subjects participating at once. As in the dictator game, subjects receive an endowment. This time, they're asked what share of the endowment they'll donate to a public pool, which is multiplied and then shared among all subjects, regardless of how much they give. As in the dictator game, researchers typically find that subjects give a reasonably large share of their endowment to the pool.

In one experiment, by John List, Robert Berrens, Alok Bohara, and Joe Kerkvliet, the research team had subjects in the control group play a game that was similar to the one we just described, asking subjects whether they would be willing to donate their twenty-dollar show-up fee to a research center on campus. In the treatment group, subjects played the same game, except that after making their decisions, some were randomly chosen and asked to announce their decision publicly, in front of all other subjects—a powerful observability manipulation! Sure enough, subjects in the treatment gave at roughly double the rates of those in the control.[15]

For more evidence, we leave the lab and head to the field (literally). In a study of fruit pickers in the UK,[16] researchers found that when they offered the fruit pickers a relative pay scheme, in which the pickers received a bonus that depended on how they performed relative to others, the fruit pickers slowed down to avoid forcing their fellow pickers to work too hard. But this only happened when the fruit pickers could see one another as they worked in the field. For some kinds of fruit, this isn't possible:

> Type 2 fruit grows on dense shrubs that are 6 to 7 feet high on average. In contrast to type 1 fruit, when picking type 2 fruit workers are unable to observe the quantity of fruit picked by workers in neighboring rows on the field-day. Hence, the physical characteristics of type 2 fruit ensure that workers cannot monitor each other on the same field day.

For type 2 fruit (we're guessing it's raspberries, but the authors didn't say), the fruit pickers failed to cooperate and ended up working way harder.

Our observability prediction ends up being pretty useful. In one of our own papers, we used this finding to promote participation in a blackout prevention program.[17] We found that people were three times (!) more likely to sign up for the program if we used sign-up sheets that revealed their identities to their neighbors than if we had them sign up using an anonymous code. Others have found similarly striking effects of observability in motivating blood donation, tax compliance, donations to national parks, and more.[18]

EXPECTATIONS

The next feature of the model that we need to check for is that people are sensitive to cues of whether they're expected to cooperate. Here's a twist on the dictator game by John List that nicely illustrates what we mean. In the standard version of the experiment, dictators have the option only of helping their partner by sharing their prize. In List's version, dictators are given the option of helping or hurting their partner: they can give their partner part of their own endowment or take part of their partner's endowment. When the decision is framed in this way, relatively few subjects give their partner any money, though, as in the regular version, few choose the most selfish option and take money away from their partner.[19] It seems subjects infer from the options available to them how they're expected to behave if they wish to be perceived as a fair or decent partner.

Another example? A variety of experiments have found that simply changing the words used to describe an experiment can influence participants' behavior. In one classic experiment, by Varda Liberman, Steven Samuels, and Lee Ross, subjects playing the ultimatum game were more giving and more likely to reject unfair offers when the game was called the cooperation game than when it was called the Wall Street game. [20] In a more recent example, by Valerio Capraro and Andrea Vanzo, subjects' generosity in a variant of the dictator game doubled when actions available were labeled as "don't steal" versus "steal" as opposed to "don't take" versus "take." Subjects were even less generous when the labels were "give" versus "don't give." One possible explanation for such results is that the labels change participants' behavior because they change participants' perception of how generous they're expected to be.[21]

If you want to see how expectations play a role in the wild, a good source of examples is the behavioral economics literature on nudges. For instance, in the early 2000s, a team of researchers led by Noah Goldstein collaborated with a hotel to test messages asking guests to reuse their towels—a minor inconvenience that saves the hotel money and helps to conserve energy and water. The researchers started out by using messages like "You can help save the environment" but found that reuse rates jumped nine percentage points when they added messages like this one: "Join your fellow guests in helping to save the environment. Almost 75 percent of guests who are asked to participate in our new resource savings program do help by using their towels more than once." The messages were even more effective when they told guests that other people who had stayed in their room had recycled.[22] Such messages, known as descriptive norms, help communicate expectations by highlighting that others cooperate in this context. A hotel guest who asked herself, "Do others cooperate in this context? Do others like me cooperate in this context?" would respond in the affirmative and reasonably conclude that she, too, was expected to cooperate.

Social psychologists and behavioral economists have uncovered plenty more clever ways to influence people's expectations and, thus, how altruistic they are. One common trick is to survey people about what they think is right (Do you feel it's fair when people avoid paying for public transportation by jumping the turnstile?) and then report the results (Over 85 percent of New Yorkers agree: avoiding your fare is unfair!). Such messages are known as injunctive norms, and they, too, can effectively communicate expectations.

There are many more examples. To check them out at the source, pick up a copy of *Nudge* by Richard Thaler and Cass Sunstein. We can't help it, though, we have to give just one more example. Back in the day, the Metropolitan Museum of Art (the Met) in New York used to be free. But, before you could enter, you had to stand in line to pick up one of those little metal badges to clip to your shirt. When you arrived at the counter, it looked just like a cashier's desk. Above the desk was a sign with suggested donations—one for regular folks and others for children, seniors, and students. You could always say you didn't want to donate that day, but the Met's default donation list set expectations to the contrary—an effective strategy.

◆

LET'S NOW TURN TO SOME INTERESTING IMPLICATIONS OF THESE FEAtures of norm enforcement. We will start by discussing some limits on what can be accomplished with norm enforcement.

LIMITS ON WHAT'S ENFORCEABLE

During the COVID-19 pandemic, most of us restricted our travel, avoided gathering with friends and family indoors, maintained a social distance of six feet from others, and wore a mask when outside our house. When we traveled across state lines, we obtained a negative COVID test result before meeting with anyone. Although many of these restrictions made their way into official local and state guidelines and, in some cases, violations were officially accompanied by a fine, most localities fiercely avoided meting out these punishments. Instead, they relied on the local population to encourage these behaviors via good old norm enforcement. Compliance was far from perfect, and it certainly varied from place to place, but still, most Americans wore masks, restricted their travel meaningfully if not quite fully, and even got the vaccine when it became available to them. Not bad, norms, not bad.

Since vaccinated individuals get and spread COVID at low rates, one would think that they would no longer be expected to wear masks and so on. Yet, it took several months for such guidelines to be lifted, and during this period, vaccinated individuals were expected to follow the guidelines along with everyone else. The official reasoning given by health authorities (that they didn't yet know whether vaccinated individuals spread the virus) was always dubious. As more data came out confirming that vaccinated individuals rarely spread COVID, they abandoned this justification but not the restrictions. Everyone was, for the time being, still expected to mask up.

The real reason why we needed vaccinated individuals to continue to wear masks was, presumably, because we still needed *un*vaccinated individuals to wear masks, and it's much easier to enforce a norm in which everyone is expected to mask up. For such a norm, third-party punishment is relatively easy. No mask? Third parties can feel free to say something, knowing others (higher-order punishers) will be on their side. Great! But requiring some people to wear masks while others didn't have to would make it harder for third-party punishers to speak up. So while it was a bit silly to ask vaccinated individuals to wear masks, it was a lot less silly than finding ourselves unable to motivate anyone to wear a mask.

This example illustrates a more general issue with norms: some things are harder to enforce via third- and higher-order punishment. Those things will be hard to sustain, and instead, we might end up enforcing something else. In the case of masks, it's no big deal. So what if some people wear masks when they don't need to?

In the next two chapters, when we delve into higher-order beliefs in earnest, we'll see that the constraints on what can be enforced can be quite meaningful. For instance, we'll see that norms must be categorical, which makes it hard to enforce certain desirable norms, like those against causing unjust harm during war. Instead, we'll be stuck enforcing norms that categorically forbid torture or the use of certain types of weapons, which are imperfect substitutes. Similarly, we might want to have a norm that encourages people to give to good charities, but we'll be unable to do so, and instead, we'll have to reward people for giving to charities, regardless of whether their money will make a meaningful difference or be squandered. We'll also see that norms are sensitive to whether harm was caused by action or inaction (did you kill those Jews or just let them die?), which, again, would ideally not be the case. For now, suffice it to say, norms are powerful, but not all powerful, and sometimes, that matters a lot.

PRACTICAL ADVICE

One of the advantages of the norm-enforcement approach is that it suggests actionable ways of motivating people to comply with norms, namely:

- **Increase observability.** If individuals can violate the norm in private, it will be difficult for third parties to punish them for it. Whenever possible, try to make the behavior in question more easily observed by others, just as we did when switching from a hotline to a sign-up sheet.

- **Eliminate plausible excuses.** Unlike the Met, the Tate Modern just places donation boxes throughout the museum, without demanding that museum-goers stand in line before entering. This leaves lots of room for plausible excuses like "I didn't see the box" or "I forgot," which makes it hard for third-party punishers to do their job. After all, they're not only concerned about whether they think a norm was violated, but are also concerned about whether others—higher-order punishers—think it was violated, and no matter how sure they might be that a museum-goer intentionally didn't contribute, they can't be sure higher-order punishers are on the same page. Whenever possible, try to eliminate such excuses. That's what the Met did. There, anyone who tries to waltz on by without donating can be third-party punished ("Hey, stand in line!") without fear that higher-order punishers would think this out of line.

- **Communicate expectations.** Since people are sensitive to cues of whether they're expected to comply with a particular norm in a particular setting, provide them with those cues when they are expected to comply. As we just saw, those cues can indicate that lots of others are abiding by the norm or that others think it's good to abide by the norm, especially if the individuals abiding by the norm are precisely the ones who tend to set or enforce it.

This checklist is useful for promoting prosocial behaviors like giving to charity, donating, conserving resources, hand washing, and antibiotic adherence. It can also be used to undermine bad norms like racism. To do so, reverse the advice: make norm violations harder to observe, provide more excuses, and send mixed messages about what's expected!

In one literature review we conducted, we found that field interventions based on these prescriptions substantially outperformed other efforts to promote altruistic behavior. Increasing observability was particularly effective, outperforming other interventions by a large margin. Communicating expectations also worked quite well—a result that has since been confirmed in other reviews. (At the time of our review, there were not enough interventions that eliminated plausible excuses for us to evaluate.) By comparison, interventions that just tried to make altruistic behavior easier or rewarded it in some material way—prize money, a mug, a T-shirt—typically led to tepid results.[23]

We already told you how we implemented some of this advice in our efforts to promote blackout prevention. Here's another example, in which we collaborated with a digital health startup to promote tuberculosis (TB) treatment adherence. TB is, unbeknownst to most Westerners, the world's deadliest infectious disease, killing over two million people each year—more people than HIV and malaria combined. Shockingly, that's despite the fact that it has a cure and has had one for seventy years.

However, TB treatment is, to put it mildly, a daunting challenge. It lasts six months or more and requires regular visits to the clinic, while the daily antibiotic doses are so powerful that they make patients feel sick. Many patients quit early, hoping they've been cured. Our goal was to keep them from doing this so they wouldn't get sick again and also so they wouldn't get others sick or develop drug-resistant TB, which is very difficult to treat.

So, here's how we approached the problem. Each day, we texted patients a reminder to take their medications. But we didn't stop there. There would have been no observability and way too many excuses: "I didn't see the text." "My phone ran out of batteries." "I lent my mom my phone." "I saw the text but forgot." So, instead, we asked patients to log in and verify that they'd taken their meds. If they didn't reply, we texted them again. And again. If, after three reminders, they still didn't reply, we had a team of supporters who would reach out and try to get them back on the wagon. Now, there was observability, thanks to those supporters, and fewer excuses. It does no good to claim you didn't see the text if you know you'll just keep getting them until you reply. We also had some opportunities to communicate expectations, by sending motivational messages like: "Together, we're kicking TB out of Kenya!" and "Already to-day, hundreds of health heros have taken their meds. Join them!"

Our intervention was quite successful. In one test of the plat-form, with 1,200 patients in Nairobi, Kenya, we reduced the pro-portion of patients who failed to complete treatment by two-thirds. In a second test, with over 15,000 patients—some 20 percent of Kenya's TB patients—we reduced this proportion by one-third.[24]

◆

SO FAR WE HAVE FOCUSED ON NORMS THAT LOOK QUITE A BIT LIKE our heavily stylized models. Reality is, of course, much richer, particularly in the ways that norms are enforced. Let's take a look at a handful of real-world norms and discuss how they are enforced.[25] Afterward, we will come back and discuss what these norms have in common with our model.

During the Golden Age of Piracy, which extended from the late seventeenth century into the early eighteenth century, pirates terrorized the merchant ships of Spain and the American colonies. Any fan of literature and movies knows that pirates cultivated a fearsome reputation. They would kill everyone aboard ships that resisted and often exercised exceptional brutality in executing their punishments. They employed theatrics to further strike fear in the merchant sailors they preyed upon. The most famous pirate of all, Blackbeard, used to hide lit fuses in his hat so that he would be surrounded by smoke. Lol.

For all the drama, pirates were shrewd businessmen, and these tactics were largely designed to minimize the need to actually engage in battle. They made a big to-do about how terrible they were, but they also showed clemency to sailors of ships that surrendered and made just as big a to-do of this. The whole point was to get merchant ships to surrender without a fight. Once they did surrender, captives were surprised to discover that pirate crews were orderly affairs. As one captive exclaimed, "At sea, they perform their duties with a great deal of order, better even than on the Ships of the Dutch East India Company; the pirates take a great deal of pride in doing things right."[26]

In strong contrast to the merchant ships of the time, where sailors were poorly paid, badly behaved, and frequently abused at the hands of their autocratic captains, pirates were well paid, well behaved, and well treated. They very, very rarely stole from one another. They made important decisions communally, often voting on the outcome. They even went to bed early; those who wished to continue drinking and gambling (on ships where gambling was allowed—it was often forbidden) were expected to go up to the deck to avoid disturbing their crewmates.

How did pirates maintain such order? It all began on land, when, upon joining the crew, each crewmember agreed to uphold the ship's code, signing it in a miniature public ceremony. The signed codes were then posted in a prominent place on the ship, often on the door to the captain's quarters (when the captain even had his own quarters—the ships were so egalitarian that captains sometimes slept with their crews). Although pirates typically made an effort to destroy these codes when captured to eliminate incriminating evidence, some have survived. Here's one, from the ship of Captain Bartholomew Roberts:

I. *Every man has a vote in affairs of moment; has equal title to the fresh provisions, or strong liquors, at any time seized, and may use them at pleasure, unless a scarcity (not an uncommon thing among them) makes it necessary, for the good of all, to vote a retrenchment.*

II. *Every man to be called fairly in turn, by list, on board of prizes because, (over and above their proper share) they were on these occasions allowed a shift of clothes: but if they defrauded the company to the value of a dollar in plate, jewels, or money, marooning was their punishment. If the robbery was only betwixt one another, they contented themselves with slitting the ears and nose of him that was guilty, and set him on shore, not in an uninhabited place, but somewhere, where he was sure to encounter hardships.*

III. *No person to game at cards or dice for money.*

IV. *The lights and candles to be put out at eight o'clock at night: if any of the crew, after that hour still remained inclined for drinking, they were to do it on the open deck.*

V. *To keep their piece, pistols, and cutlass clean and fit for service.*

VI. *No boy or woman to be allowed amongst them. If any man were to be found seducing any of the latter sex, and carried her to sea, disguised, he was to suffer death; (so that when any fell into their hands, as it chanced in the Onslow, they put a sentinel imme-diately over her to prevent ill consequences from so dangerous an instrument of division and quarrel; but then here lies the roguery; they contend who shall be sentinel, which happens generally to one of the greatest bullies, who, to secure the lady's virtue, will let none lie with her but himself.)*

VII. To desert the ship or their quarters in battle, was punished with death or marooning.

VIII. No striking one another on board, but every man's quar-rels to be ended on shore, at sword and pistol. (The quarter-master of the ship, when the parties will not come to any reconciliation, accompanies them on shore with what assistance he thinks proper, and turns the disputant back to back, at so many paces distance; at the word of command, they turn and fire immediately (or else the piece is knocked out of their hands). If both miss, they come to their cutlasses, and then he is declared the victor who draws the first blood.)

IX. No man to talk of breaking up their way of living, till each had shared one thousand pounds. If in order to this, any man should lose a limb, or become a cripple in their service, he was to have eight hundred dollars, out of the public stock, and for lesser hurts, proportionately.

X. The Captain and Quartermaster to receive two shares of a prize: the master, boatswain, and gunner, one share and a half, and other officers one and quarter.

XI. The musicians to have rest on the Sabbath Day, but the other six days and nights, none without special favour.

Codes like Captain Roberts's do not so closely resemble our norm enforcement model. Yes, they motivate compliance via (se-vere!) punishments, but the recursive logic of the norm enforcement model is absent: punishment isn't meted out by third parties moti-vated by the threat of higher-order punishment. Rather, the officers of the ship have been granted authority to mete out punishments;

sometimes, controversial cases were brought to a vote before the whole crew.[27]

The next code we'll look at is practiced by another band of outlaws: the Italian crime families of Southern Italy and the United States. *Omerta* is a code of silence that forbids members of these families from divulging information to law enforcement or outsiders—even information about competing families and even when the information would exonerate the individual being questioned. It is expected that the individual serve time, even wrongly, rather than cooperate with the police. When Vincent Gigante was brought to trial for attempting to murder mob boss Frank Costello in a botched hit, Costello refused to testify, telling police he had not recognized his assailant and that "I don't have an enemy in the world."[28] Though most inner-city drug gangs would probably not call it omerta, they abide by strict prescriptions against snitching and (as in Southern Italy) even nongang members are expected to abide.

How is omerta enforced? In 2017, Italian authorities uncovered a network of underground courts the crime families use to try mafiosos accused of breaking omerta. Punishments varied "from temporary surveillance or isolation to covering the convicted person's upper body in urine and feces, to a death sentence."[29] In the TV series *The Wire*, when drug kingpin Marlo Stanfield learns that lovable middle schooler Randy Wagstaff has provided information to the police about a murder perpetrated by Marlo's henchmen, Marlo declines to have Randy killed, but he does tell his crew to spread the word that Randy is a snitch. This proves nearly as bad. The neighborhood kids repeatedly pick fights with Randy, and one night, Molotov cocktails are tossed through the window of the apartment he lives in with his grandmother. It burns down, and his grandmother is badly injured. Randy ultimately ends up in a foster home, where his peers continue to harass him badly.

If you think omerta is unique to those outside the law, think again. The law itself practices a form of omerta known as the blue wall of silence. Police officers are expected to perjure themselves (typically by feigning ignorance) when questioned about a colleague's misdeeds. Those who don't can expect to be ostracized, harassed, or worse by their colleagues and demoted or fired by superiors. Whistleblowers are treated similarly and almost always find themselves out of a job.[30] The blue wall is a major impediment in stemming police corruption and brutality and has drawn fire from activists, especially during the Black Lives Matter movement. The Orthodox Jews, too, have a version of omerta known as *mesirah*, which forbids reporting Jews to outside authorities for many violations. Even schoolchildren (Jewish and non-Jewish alike) have prohibitions against tattling.

For our next example of a real-world norm, we turn to the lobster fisheries off the coast of Maine, which have long used a unique set of norms to self-manage their sustainable fisheries. At least since the midtwentieth century, the lobstermen have maintained strict limits on which lobsters can be sold. Those below a certain size, above a certain size, or with a V notched out of their tail to indicate they are an egg-bearing female are all returned to the sea. These norms date to the 1930s, before any laws restricting lobster catches came into effect. The laws were based on the norms, not the other way around.

In detailed studies of lobstermen, James Acheson has shown that they are, in general, a relatively orderly bunch, not too far off from pirates.[31] They are organized into gangs, each of which has a leader known as a king or kingpin—usually an older, well-respected lobsterman. The king makes sure members of the gang keep the docks clear of traps and other gear, smooths over disputes among gang members, facilitates sale of the gang's lobsters to the local dealer or cooperative, and so on. Since gangs are territorial, the king is also responsible for organizing defense of a gang's territory in the event of incursions by newcomers or neighboring gangs.

In enforcing their norms and combating incursions, the key tactic employed by lobstermen is cutting traps—severing the line that attaches a lobster trap lying on the ocean floor to the buoy at its surface. Cut traps are irretrievable, and the lobsterman whose traps are cut not only loses valuable gear but also time. If a lobsterman's traps are repeatedly and consistently cut, he or she (female lobstermen are relatively rare, but typically go by the title lobstermen as well) will soon go broke—a lesson that out-of-towners hoping to get into the business as a second career typically learn the hard way. Cutting traps is technically illegal and carries with it steep fines and potential jail time, though it is relatively difficult to catch the perpetrator. In extreme cases, lobstermen have also been known to sink one another's boats.

We now head from surf to turf, leaving behind the lobstermen on the East Coast and heading West to meet the cattlemen of Shasta County, California, who have developed a unique set of norms to ensure good relations with their neighbors. In the county's verdant rolling foothills, ranching has long been the primary economic activity. Some ranchers pay to pasture their cattle on fenced-off plots, but fences are surprisingly expensive ($10,000 per mile in materials alone), so many areas remain unfenced. These areas are still available for ranching, typically at about a third the rate of fenced lands. Of course, cattle on these lands will regularly wander off them, trespassing onto neighboring farms and ranches. There, they trample fences, eat stashed hay, and, occasionally, impregnate the cows—all costly for the neighboring farmer.

Legally, ranchers aren't always liable for these damages. It technically depends on whether the farm lies on open or closed land. On open lands, the rancher isn't liable. On closed lands, he or she is liable, not just for any damage done by cattle that trespass on neighboring farms but also in the event that a motorist hits cattle that have strayed onto a road.

Practically speaking, though, ranchers almost always respond to reports of stray cattle by quickly picking the cattle up and both literally and figuratively mending fences. If any serious damage has been done, the rancher will typically help to repair it. Here's how Robert Ellickson, who originally developed this case study, puts it:[32]

> Most rural residents are consciously committed to an overarching norm of cooperation among neighbors. In trespass situations, their salient lower-level norm, adhered to by all but a few deviants, is that an owner of livestock is responsible for the acts of his animals. Allegiance to this norm seems wholly independent of formal legal entitlements. Most cattlemen believe that a rancher should keep his animals from eating a neighbor's grass, regardless of whether the range is open or closed.

When ranchers are asked why they abide by these norms, their responses clearly indicate they've internalized the norm: "Suppose I sat down [uninvited] to a dinner your wife had cooked." It "isn't right" to get free pasturage at the expense of one's neighbors. "[My cattle] don't belong [in my neighbor's field]." A cattleman is "morally obligated to fence" to protect his neighbor's crops, even in open range.

If ranchers violate neighborly norms and fail to quickly pick their cattle up and make amends, neighbors will typically retaliate by gossiping about the rancher. This is often enough to cause the rancher to make amends, but if it fails, neighbors will sometimes take matters into their own hands, sabotaging the cattleman by moving the animals to a spot from which it will be difficult to retrieve them or even killing them. On the very, very rare occasion that a neighbor escalates his complaints to local authorities, they typically settle the dispute by calling other cattlemen and asking them to pressure the deviant into mending his ways:

> When a supervisor receives many calls from trespass victims, his first instinct is to mediate the crisis. Former supervisor Norman Wagoner's standard procedure was to assemble the ranchers in the area and advise them to put pressure on the offender or else risk the closure of the range.

The cattlemen are usually effective in this endeavor. They have good reason to be. In the past, when neighbors have felt that cattlemen haven't responded to their complaints, they have successfully petitioned to have some lands closed, which means a great deal more liability, particularly for accidents with motorists.

◆

IN ALL OF THESE REAL-LIFE CASES, COMPLIANCE WITH THE NORM IS motivated by the threat of punishment. Pirates who broke the code would be killed or marooned. Mobsters and gangsters who break omerta are killed or barely spared, like Randy Wagstaff. Lobstermen who break conservation rules or try to fish outside their territory have their traps cut. Cattlemen who don't pick up their cows find they've "wandered" into a faraway creek or disappeared entirely.

Where these examples differ is in how punishment is motivated. In the norm enforcement model and in the examples we discussed previously, punishment is motivated by second-order punishment, which is motivated by third-order punishment, and so on. That might be going on, to some extent in these examples, too. For instance, it is likely the case that lobstermen who risk their necks to cut traps on behalf of their gang are rewarded by other gang members, who can give the cutter's traps a wider berth, cede a desirable spot to the cutter, share information more openly with the cutter, lend the cutter gear, or aid the cutter with onshore chores. Those who failed to do so would likely be seen as jerks and themselves punished. Often, though, higher-order punishment is not of paramount importance, and punishment is motivated in some other ways. Here are four such ways.

1. Those who benefit from the norm compensate the punisher. Italian crime families and street gangs like Marlo's employ enforcers like Vincent Gigante—the guy who Frank Costello refused to rat on. They pay these enforcers well, and particularly loyal enforcers will rise through the ranks: Gigante would eventually take over for his boss Vito Genovese after Genovese was sent to prison. In the likely event that the enforcer is caught and imprisoned for his crimes, the enforcer's family is cared for. In *The Wire*'s final end-of-season montage, Marlo's quiet but fearsome enforcer Chris Partlow is seen in prison, chatting amicably with Wee-Bey Brice, the enforcer for Marlo's archrival, Avon Barksdale. Both are there for life, having taken the fall for their bosses in return for financial security for their families.

Even folks who don't have a formal role within the organization can be rewarded for punishing on its behalf, either informally or via bounties and rewards. This is what happens to *The Wire*'s Robin Hood character, Omar Little, who would steal drugs and money from gangs like Marlo's. When he is finally killed, it isn't by one of Marlo's enforcers but by a neighborhood kid who wishes to collect the bounty and ingratiate himself to Marlo. Of course, law enforcement also employs bounties and rewards, which motivate the public to aid in apprehending at-large criminals or enemies of the state. Throughout the United States, professional bounty hunters earn rewards for bringing in fugitives who have skipped bail. In 2019, the US promised a $1 million bounty to anyone who provided information that led to the capture of Osama bin Laden's son. In Nazi-occupied Poland, Nazis rewarded peasants who handed over or killed Jews with vodka, sugar, potatos, and oil. In Holland, the Nazis paid cash—the equivalent of about four US dollars or roughly seventy dollars in today's currency. Almost ten thousand Dutch Jews were turned over and killed.[33]

2. Institutions develop to punish norm violations. Pirates call, "All hands!" to vote on what to do about two crewmembers who got into a fistfight or a crewmember who was too inebriated to participate in a battle. The Mafia assembles in underground courts to try those who violated omerta. Lobstermen meet at the dockside eatery or tavern to resolve conflicts among gang members or to plan their defense against territorial incursions. Cattlemen presumably do the same. Humans are, after all, capable of communicating—they can get together, agree on a punishment, and work out the details of how it will be meted out. Often, they do just that.

3. The punisher is motivated to punish as a way of signaling their own commitment to the norm. When the neighborhood kids mercilessly pursue Randy Wagstaff, they are likely at least partially motivated by a desire to signal that they, themselves, would never snitch. This motive for signaling likely underlies people's eagerness to pile on and call out total strangers who say insensitive things on social media. You may recall the famous case of public relations executive Justine Sacco, who thoughtlessly tweeted an insensitive joke to her 170 followers before boarding a long-haul flight to Capetown, South Africa. One of those followers was offended and forwarded the tweet to a reporter at BuzzFeed, who retweeted it. Within hours, Sacco was public enemy number one and #hasjustinelandedyet was trending. Soon after of touching down, Sacco lost her job.

Examples of tweets critical of Sacco hint at what the tweeter gains from joining the pile on. Take this one: "In light of @JustineSacco disgusting racist tweet, I'm donating to @care today" (CARE is a nonprofit engaged in humanitarian work in sub-Saharan Africa.) Or this one: "I'm an IAC employee and I don't want @JustineSacco doing any communications on our behalf ever again. Ever." Do you think someone who tweeted such a tweet would, himself, be racist? Unlikely. If he were, he'd end up looking like quite the hypocrite.

Jillian Jordan has led a variety of studies showing such signaling motives underlie punishing. Here's one.[34] This particular experiment had two stages, a third-party punishment game followed by a trust game. Recall: in the third-party punishment game, a subject is asked whether she will pay part of her earnings to reduce the earnings of another subject who has violated the norm. In the trust game, a new subject is paired with the punisher from the third-party punishment game and asked what share of his endowment he'd like to send the first subject. Whatever he sends is multiplied, and then the first subject is asked what share she'd like to send back. Subjects who send part of their endowment to the first subject must trust that the recepient will return a share.

Jordan found that subjects were indeed more inclined to trust the first subject if that subject had punished in the third-party punishment game, sending over 33 percent more to those who punished than those who didn't. These subjects presumed that those who punished would themselves cooperate when given the opportunity. They were right: those who punished would ultimately go on to return substantially more of the funds they were sent. Even in the artificial laboratory environment, punishing a norm violation is a sign that someone can be trusted to adhere to the norm himself.

4. Punishment isn't (always) costly. Sometimes, if anything, the hard part is getting people not to punish. This seems to be true in many of the cases we've discussed. For instance, though pirate captains rarely did this, on merchant ships it was common for captains to make up fake infractions so that they could dock (and pocket) their sailors' pay. In Shasta County, neighbors who killed trespassing cattle could hardly have been sorry about the glut of beef in the deep freezer. After Frank Ellis repeatedly allowed his cattle to run roughshod over the lands of farmers and ranchers in Oak Run, California, one Oak Run resident suggested Ellis print a T-shirt saying, "Eat Ellis Beef. Everyone in Oak Run does!" Even in Sarah Mathew and Rob Boyd's studies of the Turkana, not all the punishments were costly. The fines paid by defectors were paid as animal offerings, given to those peers whom the defector had put at risk.

There are plenty of other examples that fit this story, too. The Nazis often found willing collaborators among local populations eager to plunder from the Jews they turned in. A Jewish teacher from East Poland named Stanislaw Zeminski noted in his diary, which was recovered by the Allies from the trash heap at Auschwitz, "The bodies are still warm, but people already start to write letters, asking for Jewish houses, Jewish stores, workshops or parcels of land." This was nothing new. In *Persecution and Toleration*, Noel Johnson and Mark Koyama argue that throughout the

Middle Ages, minorities like Jews were protected from persecution by local rulers because they brought in meaningful tax revenue, served important economic functions, and assisted the rulers' bureaucracies, but in tough times—during plagues or famines, for instance—those protections were lifted. The plunder they obtained from the Jews would, it was hoped, mollify local peasants, who might otherwise turn on the rulers.

Indeed, it seems it is quite common for punishments to be meted out simply by lifting restrictions on robbing and raping. In medieval England, the term *outlaw* applied to someone who was no longer protected by the law. The punishment was reserved for those who evaded punishment by other means. Outlaws could be robbed and beaten without consequence. Here's how Michael Muthukrishna and Joe Henrich describe another example, from a small-scale Fijian society:[35]

> In a subsistence-oriented Fijian community, a system involving negative indirect reciprocity—that is, tolerance of those who exploit those with a poor reputation—maintains a wide range of social norms, including those related to helping in community projects, contributing to village feasts (food sharing), and constructing one's house in a prescribed orientation. If someone violates one of these social norms, both they and their extended family fall into bad standing. If this reputation gets bad enough, after repeated violations, it is as if their reputational shield fell, and their fellow villagers can exploit them with impunity. For example, a family who violated community-wide norms by working on Sundays had some of their cooking pots and crops stolen while they were away in another village, and one of their agricultural fields was torched at night. Normally, had these acts been performed against someone in good standing, villagers would have pulled together, shared information, and tracked down the thief/arsonist. But when victims are in bad standing, villagers just shrug and let it pass.

SO, WHAT DO ALL THE NORMS WE'VE SEEN HAVE IN COMMON? IN ALL of them, norm compliance is incentivized. Exactly how? That varies. Lobstermen do things one way and cattlemen another, but at the end of the day, they make sure violations are somehow punished and punishers are somehow rewarded.

Whenever punishment requires some sort of coordination—as it did in our original model—higher-order beliefs come into play and norms become more constrained. The next two chapters are devoted to such constraints.[36]

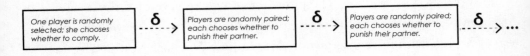

SETUP:

- There are $n \geq 2$ players.
- In the first round of the game, one player is randomly chosen to make a choice: comply or not. Complying carries some personal cost, $C > 0$.
- In the second and any subsequent round, all the players are randomly paired, and each chooses whether to punish the player they are paired with, which involves paying a cost $c > 0$ to harm that player by $h > 0$.
- With probability δ, we move from one round to the next, and with probability $1 - \delta$ the game ends.
- Players can observe everything that's happened in all prior rounds.

STRATEGY PROFILE OF INTEREST:

- Players comply in the first round.
- In the second round, players punish anyone who did not comply in the first round ("third party punishment").

- In subsequent rounds, players punish anyone who should have punished in the previous round, but didn't ("higher-order punishment").

EQUILIBRIUM CONDITIONS:

- $\delta \geq C / h$ and $\delta \geq c / h$.

INTERPRETATION:

- As in the repeated prisoner's dilemma, "observability" must be high, shirking must be punished, and punishment must itself be incentivized.
- Because the action in the first round could be anything, norms can depend on culture and context, and additional models are necessary to identify which norms are liable to emerge.
- "Higher-order beliefs"—whether a player believes others believe the norm was violated—come into play because players are motivated to punish by the threat of higher-order punishment.

CHAPTER 12

CATEGORICAL NORMS

IN THIS CHAPTER WE WILL PRESENT A NEW GAME THEORY TOOL— state-signal structures—which we'll combine with coordination games in order to explain a new puzzle: categorical norms. What do we mean by a categorical norm?

In "A Higher Form of Killing: The Secret History of Chemical and Biological Warfare," Robert Harris and Jeremy Paxman describe how "the world missed chemical warfare by . . . inches":

> The so-called Lethbridge Report drawn up for the American high command recommended soaking the island of Iwo Jima with poison gas in 1944. . . . The report was approved by the Combined Chiefs of Staff and by Admiral Nimitz, the theater commander, but when the plan went to the White House, it was returned with the comment, "All prior endorsements denied—Franklin D. Roosevelt, Commander in Chief."

Although we celebrate Roosevelt's decision to uphold the norm against chemical weapons, it came with an enormous cost—the United States lost twenty thousand men at Iwo Jima. The decision didn't even benefit the Japanese. By the time Roosevelt's generals came to him with the Lethbridge Report, Iwo Jima had been evacuated of civilians. As for the soldiers defending the island, instead of being gassed (horrible!) they were flushed out of their foxholes with grenades and flamethrowers (also horrible!). And yet, Roosevelt was willing to sacrifice twenty thousand (!!!) American lives to uphold the norm against using chemical weapons.

On its face, Roosevelt appears to have upheld the norm even when it didn't make sense to do so. Why not apply the ban against chemical weapons with more finesse, allowing their use in cases where they would cause less suffering than conventional weapons? In fact, why condition the norm on the type of weapon at all? Why not condition on the things you might think should matter, like the number of casualties or the amount of suffering caused?

The norm against using chemical weapons is an example of a categorical norm: one that depends on a categorical variable (the type of weapon used) rather than a continuous one (say, the number of civilian casualties or the amount of suffering caused).

Examples of such categorical norms abound. For instance, human rights are applied to all human beings, regardless of their degree of sentience or ability to feel pain. A chimpanzee might be more cognizant than a newborn human or more pain sensitive than a comatose one and yet is granted far fewer rights. Why do we not simply grant rights proportional to sentience or ability to feel pain?

Another example? Many of us also consider rights inviolable. We consider it egregious to violate someone's rights, say, by torturing them, regardless of the social benefits—even if it's to discover the location of a ticking time bomb that would kill a gaggle of schoolchildren. Why not consider rights violable, depending on the social gains from violating them? Why not weigh the harms against the benefits when deciding whether or not to violate someone's rights?

Less admirable norms are also often categorical. The Jim Crow South's norm that Black people should give up their seats to white people did not require that anyone with a darker skin tone give up their seat to anyone with a lighter skin tone. Rather, race was defined categorically, based on the infamous one-drop rule, which asserted that a person with even one Black ancestor was considered

Black. Why didn't Southerners have a more continuous basis for discriminating according to skin tone?

In this chapter we will try to address questions like these. But first, we need to introduce that new game theory tool.[1]

STATE-SIGNAL STRUCTURES

Our new tool has two parts. The first is states. We've seen states a few times already. In the hawk-dove game, for instance, we had two states: player 1 arrived first and player 2 arrived first. We also saw states in our chapter on evidence games. There we had two states, high and low. Although it's a bit pedantic, one could represent a peacock's fitness levels as states, the two states being fit and unfit.

Where there are states, there are priors. Priors just indicate the odds that a state might occur before players learn anything about it.[2] For instance, each player might have an equal likelihood of arriving first. Or, maybe, player 1 arrives first 80 percent of the time. Maybe peacocks are fit half the time, maybe they are fit only 10 percent of the time.

Signals capture what information players might have about the state. These are unlike the signals in the costly signaling model in that they are not generated by a player but by nature. For instance, each player might always get a signal that exactly corresponds to the state. That's what we assumed in our analysis of the hawk-dove game when we assumed that players always knew who arrived first, with no mistakes. That's how we would represent signals that are perfectly informative and have no noise. Of course, players' signals are often noisy (and this is often when things get particularly interesting). Players might, for instance, get a signal that corresponds to the state 90 percent of the time but get the wrong signal 10 percent of the time.

A state-signal structure is then just the combination of states (including their priors) with signals (including the appropriate description of how signals are generated in each state).

ADDING A COORDINATION GAME

Next, we're going to model the decision of two players who represent observers deciding whether to punish a rogue actor. Maybe they're two world powers deciding whether to impose economic sanctions on a murderous regime that has unleashed violence on its own citizens. We'll model this using a very, very simple game known as the coordination game.

In the coordination game, each player chooses between two actions, which are usually labeled A and B, though we will label them sanction and don't sanction. The payoffs of the game are such that players prefer to sanction if and only if the other is sanctioning. (We'll justify this assumption momentarily.) The payoffs can be summarized by a single parameter, p, which represents how confident a player needs to be that the other is sanctioning for her to prefer sanctioning. If $p = .78$, then a player would need to be 78 percent confident that the other is sanctioning before she prefers to sanction herself. If she puts equal odds on the other sanctioning, then that wouldn't be enough. In that sense, not sanctioning is the safer bet. Economists have a technical term for the action that's a safer bet in the coordination game. They say it's *risk dominant*. Whenever $p > \frac{1}{2}$ not sanctioning risk dominates sanctioning. When $p < \frac{1}{2}$, it's sanctioning that's risk dominant.

The coordination game is really intended as a simple proxy for any game that has a coordination element, which includes any game with more than one equilibrium. We have already encountered several such games. For instance, in the hawk-dove game, there were two equilibria, (hawk, dove) and (dove, hawk), and it was imperative that players coordinate on which one they were playing: a player wouldn't want to play hawk if he is expected to play dove and vice versa. There was also a coordination element to the repeated prisoner's dilemma: players could be playing ALLD, grim trigger, or a host of other strategies, and if one player plays ALLD when the other expects her to play grim trigger, then, hoo boy, was that a missed opportunity. The same was true in the norm enforcement game, where it would really suck to be the one guy who tries to enforce one norm when everyone else expects another, or for that matter, to be the only one insisting on enforcing sanctions on a rogue regime. In all of these games, coordination matters. The coordination game is just a simple way for us to hone in on this element of those games and others like them.

To this simple coordination game, we're going to append a state-signal structure. What we mean by this is that, first, the state is determined according to the priors and then players get their signals according to the distribution for that state (a game theorist might say that nature determines the state and sends the signals). Only after players get their signals do they play the coordination game.

The players will be able to condition their action in the coordination game on their signal, but they don't have to. In fact, the state and signal won't directly affect the coordination game, so the only reason players would condition their behavior on their signal is because they expect others to and need to stay coordinated.

Why make it so the state and signal don't directly alter the coordination game payoffs? Because when states or signals directly influence players' payoffs, it's not so surprising if players condition their behavior on them. However, we're particularly interested in understanding the kinds of factors that can influence coordinated action even if those factors don't directly matter.

This is similar to what we saw in the hawk-dove game when discussing uncorrelated asymmetries. Recall that there were factors—such as who arrived first—that affected behavior in the game even though they didn't affect the value of the contested resource or the likelihood of winning a fight. In that chapter, we established that uncorrelated asymmetries can influence behavior. In this chapter (and the next), we're asking: What can act as an uncorrelated asymmetry?

We're also asking this question when it comes to dyadic relationships and norms. What kinds of transgressions can lead to a breakup in cooperation? What kinds of norms are enforceable? In both these cases, it also makes sense to focus on information that's independent of payoffs. In the repeated prisoner's dilemma and the norm enforcement game, past defections, norm violations, and failure to punish are in the past, so they don't directly affect payoffs from subsequent behavior. Yet, sometimes future behavior does depend on them. When is that possible? That's the question these two chapters explore.

In this chapter, we're going to focus on one particular feature of states and signals that makes it more or less possible to condition on them: Is the signal continuous or discrete? We will see that it is much harder to coordinate on continuous signals than discrete ones. Let's model that now.

CONTINUOUS VERSUS DISCRETE SIGNALS

The first state-signal structure we're going to consider will have continuous states and signals. Specifically, we'll allow the state to be any number between 0 and 1. This might, for instance, represent the share of the population killed by the murderous regime that's being observed by our two players. We also need priors. For now, we will simply assume that all states are equally likely.

Our two players don't directly observe the state; that is, they don't see the actual share of civilians killed by their murderous despot. All they see is their signals. We're going to assume that each player independently gets a signal that's equally likely to be within some small range of the truth (say, .01).[3] These signals represent the observers' own estimate of the share of civilians killed by the murderous despot.

After players see their signal, they play the coordination game, independently deciding whether to sanction, and getting payoffs as summarized by p.

We're particularly interested in strategies in which players switch from not sanctioning when their signals are low to sanctioning when they get a signal above a certain threshold (let's say the threshold is .05). The question is: Is such a threshold strategy an equilibrium?

The answer is no. To see this, put yourself in a player's shoes and imagine you get some arbitrary signal. Maybe it's .179. Now, you must evaluate the odds that the other player will also get a signal above .05 and, hence, will be sanctioning. These odds are 100 percent, since signals can only deviate from the truth by .01 and, thus, from each other by .02. So, it makes sense to sanction as prescribed. No problems. Yet.

A problem arises, though, for signals close to the threshold, like .050001. Now you are basically 50 percent sure that the other got a signal below the threshold and, therefore, won't be sanctioning. If not sanctioning is risk dominant, this means you would rather play it safe and stick with "don't sanction'" even though you are supposed to be sanctioning above the threshold. That's a problem. We have found an instance where you would want to deviate from the proposed equilibrium strategy. This proves that the proposed strategy isn't in fact an equilibrium.

Notice that if sanctioning were the risk dominant action, an analogous argument could be made for signals just *below* the threshold. Also, the argument didn't really depend on the threshold being .05. In fact, no threshold strategy will be an equilibrium (except in the case where p is *exactly* .5, a case that will never happen in practice).

Let's contrast this result with what happens when the states and signals are discrete. Imagine the state can only take two values: 0 or 1. This might represent the more categorical question: Did the despot use chemical weapons?—with 1 = yes, 0 = no. We can again assume that each state is equally likely. Once again, players don't directly observe the state, only their signals. We'll assume that each player independently gets a signal that corresponds to the state (1 in state 1, 0 in state 0) with probability 1 - ε, where ε is an error term (say, .1). This means observers' weapons inspectors usually correctly identify whether chemical weapons are used, but sometimes (10 percent of the time), they screw up.

Now, we'll consider the strategy in which players switch from playing don't sanction when they see 0 to sanction when they see 1. Is this an equilibrium? Yes, so long as ε isn't too big.[4] That's because when players see a 1, they will be sufficiently (more than p) confident the other did as well, and when they see a 0, they will also be sufficiently confident the other saw a 0. It's possible to condition on categorical signals.

This result is somewhat paradoxical. In the continuous state-signal structure, players' signals are, in a sense, more refined and informative: players can learn whether the rogue actor did a lot of evil or just a little. And we can make players' signals very precise—arbitrarily precise. And yet, so long as there is a tiny amount of noise, threshold strategies can't be used in equilibrium. Meanwhile, in the discrete case, players' signals are blunt—they have just one bit of information—and this signal can be fairly noisy. Doesn't matter. Nash says we can't use the continuous signal but can use the discrete one. In coordination games, less is more.

UNRAVELING

Should we really care what Nash says? Will the fact that the threshold strategy—sanctioning for signals above, say, .05—is technically not a Nash equilibrium prevent it from being used in practice? Just because there was a beneficial deviation in a very specific instance very close to the threshold? Yes.

One way to see this is to imagine players are able to use induction—applying the same logic over and over in their heads. They'll realize that if someone who receives a signal within, say, .01 of the proposed threshold deviates to the risk dominant action, this means that anyone who gets a signal within .02 of the threshold should also deviate. But then, of course, anyone who gets a signal within .03 of the threshold should deviate and so on, until no one would ever want to sanction. That's the inductive argument for why a deviation close to the threshold would cause people to deviate much further away.

But maybe people aren't great at induction, or maybe they don't expect others to be? That won't save the threshold strategy either, so long as people learn or behavior evolves. Imagine that players simply respond the way they are prescribed, except sometimes they experiment, and notice that they can get higher payoffs by deviating. Then, eventually, in the small region where a deviation is beneficial, enough players will learn to deviate. Now it will be beneficial to deviate just beyond that, and soon enough, players will deviate there, too. This process will continue, gradually unraveling the threshold strategy until, eventually, there is no threshold at all. It could take a while, but still, no threshold strategy will be observed in the long run.

KEY ASSUMPTIONS

So far, we have presented an example where it was impossible to condition on continuous signals. A theorist might then ask: What does this result depend on? Here are the three key assumptions:

1. **Signals are noisy.** If there were no noise, whenever one player got a signal above the threshold, she would *know* the other got the same signal and, thus, that his signal is also above the threshold. That's true regardless of how close to the threshold her signal was. In this case, there is no problem sustaining continuous norms.
2. **Signals are private.** If players observe a public signal, or can communicate and verify their signals, then there is no problem having a continuous norm. Whenever a player gets a signal indicating the threshold was breached, she will know, or can ensure, the other player got such a signal as well.
3. **Coordination matters.** If players don't much care what others are doing or thinking but are instead motivated to take different actions in different states (for instance, to accept fitter peacocks), then there is no problem conditioning on a continuous variable. In fact, the more fine-grained their

signals—the more continuous—the better. This is true, for instance, in costly signaling games, where the sender could take on a continuum of types (the peacock's fitness could take any number from 0 to 1), and there could be a continuum of possible signals for him to send (the peacock could grow a tail that's anywhere from 0 to 1 meter long). It's no problem for the receiver to condition on the signal in that game.

Here are some things this result doesn't seem to depend on. First, the precise amount of noise doesn't matter. We can make it arbitrarily small, but as long as it is not zero, we will get the same result: players can't condition on a continuous signal.

Also, the precise distribution chosen for the states and signals doesn't matter. If instead of choosing a uniform distribution for states and signals, we'd chosen, say, a normal distribution, the math would have been different, but it would still be harder to get a threshold equilibrium in the continuous case than in the corresponding discrete one.

Finally, to get a threshold equilibrium, we didn't need it to be the case that the state and signal only took two values. We could let the state and signal take on more values. However, the more values they can take (and the closer to continuous they look), the harder it will be for there to be a threshold equilibrium.

It's also worth emphasizing one final thing that sometimes trips folks up. It might be tempting to think we can solve the problem with continuous signals by chunking them into categories, for instance, saying that when players get a signal above .05, they should categorize this as a 1, and when they get a signal below .05, they should categorize this as a 0. Unfortunately, this doesn't solve the problem. That's because players still have access to the fine-grained information. They can still get a signal .050001, and if they do, they will still be tempted to deviate regardless of whether this is chunked and interpreted as a 1. To enable coordinated action, players must actually get less fine-grained information.

BACK TO OUR PUZZLES

Let's go back to the norm against chemical weapons and see how our model helps us understand it.

We might in an ideal world enforce sanctions against a country that kills more than, say, 5 percent of its civilians, instead of relying on the type of weapon they use as an imperfect proxy for how much they harm or how much suffering they are causing. However, in reality, three factors prevent us from being able to do so. First, we typically don't observe the share of civilians killed with certainty. Rather, we get signals of this share that are imperfect; each observing country relies on its own intelligence agencies to guess the actual number of civilians impacted. Second, countries' signals are at least somewhat private. Each relies on its own intelligence agencies and can't easily share top secret intelligence. Each also has its own sets of priorities and, knowing this, would be unlikely to take another's word at face value. Third, there is a need to coordinate. No country wants to be the only one to enforce sanctions. Indeed, that's probably a recipe for getting into a trade war or for sanctions that don't hurt anyone but the sanctioner.

For an extreme example that highlights the presence of all three of these problems, consider the recent case of Syria's Bashar al-Assad, whose regime's brutal crackdown on political dissidents in 2011 kicked off a civil war that rapidly accumulated tens and then hundreds (!) of thousands of casualties but elicited no serious threat of military interference by foreign powers until it used chemical weapons on its citizens.

Let's check that our three key assumptions are met in this example. First, are observing nations' signals noisy? Absolutely. Each relies on hazy intelligence, and even now, the exact number of casualties is far from known. Second, are signals private? Yes. Each nation relies on its own intelligence, which it cannot easily verify to others. And some nations, like the United States and Russia, clearly have opposing interests and don't take each other's word. Third, do sanctions require coordination? Absolutely. Assad's regime was subjected to sanctions by nearly all the world's major economies and yet managed to stay afloat because Russia refused to sign on.[5]

The logic for our other puzzles is similar. Why do human rights apply categorically, to all members of the species *Homo sapiens*, but not continuously, based on a being's ability to feel pain or degree of sentience? Presumably because we won't always agree on which organisms can feel sufficient pain or are sufficiently sentient, but we can agree on who is a living, breathing member of *Homo sapiens* (indeed, we made this easier by killing off our closest relatives). And human rights, contrary to what many might purport, only exist to the extent that we believe in them and are willing to enforce them—something that requires coordination.

It's a similar story when it comes to why human rights are treated as inviolable. The norm to never torture is easier to enforce than one that says: "Don't torture unless a sufficient number of lives are likely to be saved." It's hard to agree on when the sufficient number barrier has been breached.

What about the infamous one-drop rule of the Jim Crow South? This rule made it easier for segregationists to agree on when their racist norms were violated. If someone had any discernible African ancestry, they were considered Black. If they then didn't give up their seat to someone who had no noticeable African ancestry, they would have violated the norm and would be subject to sanctioning. People who sanctioned them would be "doing their job." If a more continuous measure of race were used, like whether someone's skin is sufficiently dark to be identified as Black, there would have been more disagreement on when to enforce. Is this particular darker-skinned individual violating the norm by not getting up for this lighter-skinned one? It might, at least in some circumstances, have been hard to say, and certainly to be sure that others agree, and this could cause the norm to eventually unravel (something we might have wanted, but those who created and pushed these norms didn't). The one-drop rule is just one example of the many artificial boundaries that groups use to better coordinate on who is deserving of better or worse treatment.[6]

EVIDENCE

As usual, we will present evidence for the above model by looking at what happens when the assumptions more or less hold—what we have been calling comparative statics.

We'll make two comparative static predictions. The first focuses on how important coordination is. If our story is right, we should see less reliance on categorical distinctions when coordination is playing less of a role, when signals are perfect, or when information is shared (or easily shareable). Do we?

In Jane Austen's *Pride and Prejudice*, we witness firsthand what goes through Elizabeth's mind as she considers whether to marry her suitor, Darcy. She first meets Darcy at the Meryton Ball in the book's opening, when she and the other ladies can't help but notice how tall (a continuous variable), handsome (another continuous variable), and wealthy (also continuous) he is, though it soon becomes clear that Darcy is also proud (continuous), and Elizabeth loses interest. It takes the better part of 122,000 words, during which Darcy works hard to become less arrogant, before Elizabeth is finally won over. On a surprise visit to Darcy's estate, she learns that he is sweet tempered (continuous), good-hearted (continuous), and thoughtful (continuous).

The decision of who to date and marry is largely, though not entirely, a private one, requiring little coordination with others. If Elizabeth likes—nay, loves—Darcy then she can marry him. It's up to her to decide. Indeed, in this case, we see that Elizabeth's decision is based on continuous variables. Of course, she's hardly alone. Anyone who has ever dated has discriminated based on continuous variables like height, attractiveness, intelligence, age, and so on.

There is one categorical variable that looms large in Elizabeth's decision, though, and that's nobility. Elizabeth and Darcy would not have even glanced at each other were it not for their noble status, which makes them eligible to attend the Meryton Ball and to marry each other. It doesn't matter that Elizabeth's family is relatively poor. So long as she is of noble birth, she is eligible for the match—a categorical distinction if ever there were one. Notice, though, that discriminating against non-nobles isn't something that Elizabeth does of her own accord. It's part of a larger, coordinated effort by a tight cadre of elite, noble families. Failing to marry someone within this noble cadre was grounds for being sanctioned or excluded by other nobles, as cash-strapped English families who married their sons to the daughters of wealthy American business tycoons often discovered.

An analogous story rears its head in another exclusive culture, the Orthodox Jews. The most fervent Orthodox Jews within a particular sect are sometimes referred to as God-fearing. They're the sort who would never take advantage of loopholes and who stick to the spirit of the law, not just its letter. They also go out of their way to perform more mitzvahs—good deeds done out of piety—than is strictly necessary. The more they do, the more God-fearing they are—a continuous variable.

When choosing a husband, the bride-to-be and her family will attend to a host of continuous variables, such as just how God-fearing the suitor and his family are and how generous they are in their donations to the synagogue, as well as some of the same things Elizabeth might have attended to, like how kind and handsome the groom is. However, when it comes to determining who's observant—who's in the in-group and who's an outsider—these continuous variables fall by the wayside, and instead, the community relies on categorical distinctions, like does the individual in question keep kosher, keep the Sabbath, wear modest clothes (defined categorically as covering the collar bone, elbows, and knees), and so on. These are all yes or no questions; there is no more or less.

Here are two more examples in which people do make use of continuous variables when there is less of a need for coordination. The first is one you've likely encountered: hiring or admissions, where those assessing applicants attend to continuous variables like intelligence, scores on exams, sociability, and so on in much the same way as Elizabeth and her peers did when assessing suitors. The second: often a person or institution has unilateral authority to sanction, without the need to coordinate with others. The US Environmental Protection Agency (EPA), for instance, has the authority to regulate and fine polluters, by itself, without having to coordinate this effort in any way. Often, its regulations depend on the amount of a pollutant like arsenic that is detected in the water or air outside an industrial facility. The regulations typically allow for some amount of the pollutant. In the case of arsenic, for instance, it's up to ten parts per billion. If the EPA measures more arsenic than this, it levies a fine. In doing so, the EPA is relying on a threshold (ten parts per billion) in a continuous variable (the amount of arsenic in the nearby waterway), but since there's no need for the EPA to coordinate with others when enforcing its laws, this doesn't create any problems.

Let's now turn to a second comparative static, which focuses on uncertainty and private information. Aside from relying on coordination, our model also relies on private information: everyone must get at least somewhat distinct signals for continuous norms to be a problem. When public cues are available, so that everyone sees exactly the same thing, do we see more continuous norms? Here are two examples which suggest that the answer is yes. The first example is one that those of us from the United States encounter on a regular basis: tipping. Tipping is a norm that relies upon a threshold in a continuous variable: you could tip 22.9 percent, 17.2 percent, 15.4 percent, 13.7 percent, or any other number, but if you tip less than 15 percent (or, in some places, 20 percent), you're viewed as miserly and risk a tongue-lashing. How come we can rely on a continuous variable for tipping norms when we can't for most others? One reason might be that tipping is just a means of impressing the waitstaff or your fellow patron—more signaling than coordination. But another possibility: people don't get independent signals of how much you tipped. Rather, when they get a cue at all, observers get exactly the same signal—they look at the very same check—and anyone can look at this one and only signal at any point to verify whether the person has violated the tipping norm.

Another famous case where continuous norms were used was in the case of sharecropping—the practice of landlords and their tenant farmers splitting the food grown by the tenant on the landlord's land. The share that the landlord got was a continuous variable: it could be 22 percent, 38 percent, 64 percent, and so on. In most places, though, it was 50 percent, and if the landlord tried to demand more, his tenants could rise up against him (see Chapter 5). As with tipping, it's perhaps surprising that sharecroppers were able to employ a threshold norm, but again, we see that the tenant and the landlord's signals are shared. The food is out there in the fields for everyone to see—it's not like the tenant has anywhere to hide it.

UNRAVELING IN PRACTICE

Our model suggests that even minor breaches of a categorical norm could lead the norm to unravel. The rapid collapse of the norm against bombing cities from aircraft that held for the first months of World War II provides an illustrative example of such unraveling. On September 1, 1939, when the United States was still a neutral party in the war, President Franklin D. Roosevelt issued an appeal to both sides urging them to withhold from bombing cities.

> I am therefore addressing this urgent appeal to every government which may be engaged in hostilities publicly to affirm its determination that its armed forces shall in no event, and under no circumstances, undertake the bombardment from the air of civilian populations or of unfortified cities, upon the understanding that these same rules of warfare will be scrupulously observed by all of their opponents.

France and the UK officially agreed to Roosevelt's request, and while Hitler did not formally reply, he did order that the Luftwaffe refrain from targeting civilians. Both sides targeted industrial and military sites in cities, and since bombing was hardly a precise affair, especially at night, there were often civilian casualties and homes, stores, and churches were frequently destroyed. However, roads, bridges, food storehouses, and the like were not intentionally targeted. At first.

In July 1940, the Luftwaffe launched a concerted effort to destroy the Royal Air Force in what we now call the Battle of Britain. Over the coming weeks, they bombed the bejesus out of British targets, but always, their targets were military: airfields, radar stations, factories, and so on. Then, on August 24, during a massive bombing campaign targeting docks and factories on the outskirts of London, a small number of bombs fell on Central London, likely dropped by accident by German bombers who had missed their targets. Churchill responded by ordering a bombing raid on Berlin's Tempelhof Airport, right in the center of the German capital. Despite minimal casualties, Hitler immediately responded by ordering that the Luftwaffe shift its attention from bombing airfields to bombing cities, with the aim of disrupting Britain's economy and food supply. Britain responded in kind. By the time they areabombed Mannheim on December 15 and 16, the norm against bombing cities had turned to rubble.[7]

Unraveling of this sort is presumably why lawyers, politicians, and schoolmasters are so often worried about slippery slopes. Albeit with less deadly consequences.

It might also be why it's common for lawyers to purposely select sympathetic test cases, which breach a category barrier and lead a categorical norm to unravel. Ruth Bader Ginsburg was highly effective in her use of this tactic, choosing to represent male plaintiffs to whom the court would be sympathetic in cases like *Frontiero v. Richardson* and *Weinberger v. Wiesenfeld*, in which she challenged gender discrimination laws that, on average, harmed women. Once the court ruled in favor of those it was sympathetic with, this set a precedent that applied more broadly, and the categorical norm crumbled.

BONUS APPLICATION: (IN)EFFICIENT GIVING

In Chapters 10 and 11, we learned how repeated games and norm enforcement help to explain some features of altruism. But in the introduction to this book, we discussed some quirky features of altruism that haven't yet been explained. One was that while we are quite willing to give, we often aren't so sensitive to whether our gifts are used in effective ways. In a famous study that we briefly referenced in the introduction, subjects were asked how much they'd donate toward nets that would save birds from being killed by wind turbines. The researchers changed just one thing: the number of birds that would be saved by the nets. It was either 2,000, 20,000, or 200,000. Despite the tenfold and hundredfold increase in the benefit of the nets, and thus a tenfold and hundredfold in the benefit of giving to this cause, subjects' responses were virtually identical across all three treatments. Such insensitivity to the impact of giving has important practical implications. Empirical studies suggest it is not uncommon for there to be charities devoted to the same cause that differ one hundredfold in their impact. If people would only divert their gifts to the most effective charities, their money would go so much further.

This quirk of inefficient giving maps quite neatly to the model in this chapter. (There were other quirks we referenced in the introduction as well, for instance: the same people who would help out if asked will try to avoid being asked, and the same people who would help out if they know help is needed do their best to remain strategically ignorant of this need. These will be addressed in the next chapter.)

Here's how our games map onto inefficient giving. Our analysis of the discrete case showed us that it's relatively easy to have a norm that encourages philanthropy. When someone gives to a charity, that's a discrete action, and as we've already seen, we can coordinate to provide those who give with the necessary kudos and other social rewards to motivate such giving.

Such a norm, though, is silent on how effective the charity in question must be. Would it be possible to incorporate effectiveness somehow? Could we have a norm that says only give to a charity that's sufficiently efficient? This is harder.

For starters, it's hard to agree on exactly what counts as effective. Indeed, there is a hot debate among charity evaluators over which measures of effectiveness are best and exactly how to measure them. Some evaluators use the ratio of program expenses, which measures the percentage of funds donated that are actually devoted to the cause in question and not spent on salaries, plane tickets, rent, and other overhead. Others use quality-adjusted life years (QALYs), which tries to determine how much a charity improves people's quality of life. Others use disability-adjusted life years (DALYs). And so on. Each such measure has its benefits and downsides, and each has its own measurement challenges. What exactly should count toward overhead? Just how many QALYs are associated with each malaria net? Even professional evaluators come to slightly and sometimes very different conclusions about the effectiveness of the same charity.

Moreover, each of these measures is continuous: the ratio of program expenses can be any number between 0 and 100, and QALYs per dollar can, in theory at least, be any positive number. Thus, the situation maps quite closely to the one we've modeled: we'd like to reward only those who give to charities whose efficacy is above a certain threshold, but since we can't quite agree on a measure of efficacy, and efficacy is continuous, we can't easily do this.

Is this explanation for inefficient giving right? Our friend Bethany Burum designed and ran some experiments to verify that charitable giving is in fact insensitive to efficacy and, more importantly, that this insensitivity is primarily driven by the way social rewards work.[8]

In Burum's first study, one group of subjects was asked what proportion of their annual income they'd pay to save one person from starving. Their average response was about one-tenth. Then, another group was asked what proportion of their income they'd pay to save five people. How much more were they willing to pay? You guessed it. Even though five times as many lives would be saved, they respond virtually identically to the first group. People's willingness to pay was insensitive to the impact it would have. In the second study, Burum gave subjects a small bonus and asked them if they'd be willing to donate part of that bonus to a charity. She told some subjects their donation wouldn't be matched. Others were told their donation would be matched 1 to 1. Others 2 to 1. All the way up to a multiplier of 10 to 1. That is, a dollar donated by these subjects would go ten times as far as those in her first treatment. Subjects were again quite insensitive to impact: they gave the same on average (about a third of their bonus), regardless of whether the multiplier was 1×, 2×, or 10×.

But what happens when she changed the setting a bit to one where norms of giving played less of a role? One way Burum did this was by having subjects ponder giving toward kin instead of strangers. Of course, we all expect people to feel more altruistic in this case. But should our altruism be more sensitive to the *numbers* who would be helped? Yes, if you think our altruism toward our kin can be sensitive to impact because altruism toward kin is driven less by norms and more by shared genes. And that's what Burum found. When kin were the imagined people being saved, subjects' willingness to spend depended a lot more on the number to be saved: subjects said they'd pay about a third of their annual income to save one family member but about half to save five family members. That's not five times as much, but it's a big jump.

Burum verified the crucial role social rewards were playing in two additional ways. One was by swapping the charity decision with a savings decision but still varying the multiplier from 0 to 10. Savings decisions, unlike charity decisions, are personal, not social. For such decisions, what others think and know is less of a concern. And, in fact, subjects saved much more when the multiplier was higher—about twice as much for the highest multiplier compared to the lowest.

One final way Burum verified the role of social rewards: she asked other subjects to evaluate those who gave and didn't give, varying the multipliers the would-be givers were said to have faced. The evaluators were given the option to put their money where their mouth was: paying some money to reward those who did or didn't give. Bethany found that these evaluators were sensitive to whether the person they were evaluating gave but not to the multiplier they faced when making this choice: they rewarded people who gave a lot more, regardless of what multiplier their gift would have been matched by.

◆

IN THIS CHAPTER, WE USED STATE-SIGNAL STRUCTURES TO UNCOVER a surprising result: when coordinating, players cannot condition on continuous private signals if they have even the tiniest bit of noise. This helped us understand why, when it comes to norms, altruism, and presumably other contexts where coordination plays an important role, people tend to rely more on categorical distinctions like species membership or whether someone keeps kosher, rather than on the continuous information that one might have thought they'd actually care more about, like ability to feel pain or how pious someone is.

In the next chapter, we'll use state-signal structures and coordination games to explore the influence of higher-order beliefs on our sense of altruism and morality.

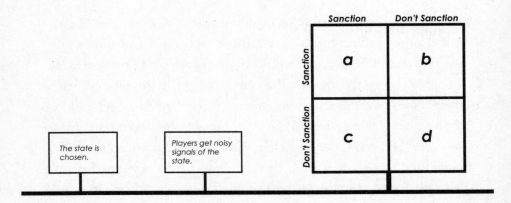

CONTINUOUS CASE

SETUP:

- The state of the world is chosen uniformly from [0,1].
- Players signals are drawn independently and uniformly from within ε of the true state, for some ε ≥ 0.
- Players then simultaneously choose their action in the coordination game, which is presented in the payoff matrix. The parameter $p = (d - b) / (a - c + d - b)$ indicates how confident players must be that the other is sanctioning before they would want to sanction. Payoffs are not directly affected by the state or the signals.

STRATEGY PROFILE OF INTEREST:

- Threshold strategy: players sanction if and only if they receive a signal above some threshold value, s^*.

EQUILIBRIUM CONDITIONS:

- Whenever $1 > s^* > 0$, threshold strategies are never equilibria, except when p is exactly .5 or $ε = 0$.

INTERPRETATION:

- It is impossible to sustain norms that condition on continuous information, except when the information is error free.
- Had we allowed players' to share their signals, or receive the same signal, this, too, would allow players to condition on the information in equilibrium.

DISCRETE CASE

SETUP:

- The state is 1 with some known probability, μ, and 0 otherwise.
- Each player independently gets signal 0 or 1, where the signal corresponds with the state with probability $1 - \varepsilon$.

STRATEGY PROFILE OF INTEREST:

- Conditional strategy: players sanction if and only if they get signal 1.

EQUILIBRIUM CONDITIONS:

- So long as ε is small enough, the conditional strategy is an equilibrium. (The precise condition is: $[(1 - \varepsilon)^2 \mu + \varepsilon^2(1 - \mu)] / [(1 - \varepsilon)\mu + \varepsilon(1 - \mu)] \geq p \geq 1 - [(1 - \varepsilon)^2 (1 - \mu) + \varepsilon^2\mu] / [(1 - \varepsilon)(1 - \mu) + \varepsilon\mu]$.)

INTERPRETATION:

- It is possible to condition on categorical information so long as the error rate isn't too high.

CHAPTER 13

HIGHER-ORDER BELIEFS

LIKE THE LAST CHAPTER, THIS CHAPTER IS GOING TO MODEL THE KIND of information that can and cannot influence coordinated action. Except, now, instead of focusing on the importance of categorical distinctions, we are going to focus on the role of higher-order beliefs—what you believe others believe, what you believe others believe you believe, and so on.[1]

But, first, some puzzles:

Ceremonies and symbolic gestures. In 2000, Bill Clinton hosted Yasser Arafat and Ehud Barak at Camp David to attempt to negotiate a peace between the Palestinians and the Israelis. In an amusing scene caught on video by the gaggle of reporters who assembled to witness the historic event, Arafat and Barak tussle over who should go through the door first, each insisting the other do so as a sign of respect and refusing to back down until Clinton, laughing loudly, opens the second door so that both can enter into the building simultaneously.[2] Why would the two care about such a minor gesture?

In fact, our lives are replete with gestures to which we pay seemingly undue attention. Why do mere handshakes between leaders (Arafat and Rabin, Trump and Kim Jong-un) make the news? Why do we fly flags? Why do we wait with bated breath for our partner to say "I love you" for the first time or for a first kiss? Why do we say please and thank you reflexively, barely noticing these words unless they are absent? Why do we attend so closely to whether others make eye contact? What information do these gestures convey? What do we learn from the handshake, from I love you, from please and thank you, or from eye contact? Nothing, it seems. Yet even though they don't communicate any novel information, they carry much weight.

The same holds for the myriad ceremonies that punctuate our lives—baptisms, bar mitzvahs, graduations, weddings, inaugurations, and coronations. What purpose do these serve? It's not like our families learn much about us at our bar mitzvahs, graduations, or weddings. They know how old we are. They also know whether we're on track to finish high school long before June of our senior year. And they know we love our partner all too well, since we've, for years, subjected them to a parade of too-cute photos on Facebook (see Chapter 8). Why do nations come to a standstill—and shell out princely sums—for parades and speeches when it's time to hand the reins to a new king, president, or prime minister? What do we learn from these overwrought celebrations? They, too, carry great weight even though they rarely, if ever, communicate any novel information.

The omission-commission distinction. In 1942, the MV *Struma*, a seventy-four-year-old luxury yacht turned livestock shuttle, took to the sea for the last time. For three months, the boat had sat in port in Istanbul, with a failed engine. Finally, the authorities got sick of her, attached her to a tug, towed her out to the open ocean, and left her there. A day later, a Russian submarine sank her. There were 781 Jewish refugees and 10 crew members onboard. One survived.[3]

The Turkish and English authorities that had the *Struma* towed to sea knew all too well what would befall the stranded boat. Soviet subs had standing orders to torpedo neutral ships like the *Struma*! Yet, while they were willing to knowingly let harm befall the Jewish refugees, the Brits were unwilling to torpedo the boat themselves. Their behavior illustrates a quirk of our morality known as the omission-commission distinction—that we tend to be more comfortable knowingly enabling harm via inaction (by omission, for instance letting the Russians torpedo the boat) than via action (by commission, for instance torpedoing it ourselves).

Quirks of our morality like the omission-commission distinction are often expressed using clever thought experiments designed by philosophers, known as trolley problems. In these, an out-of-control trolley hurtles down the track, and subjects are forced to choose between two pretty awful outcomes. Would you pull on a lever that switches the track, which would result in the trolley killing just one person instead of five (a commission)? Would you push a man with a heavy backpack, who just happens to be nearby, in front of the trolley to stop it from killing five others farther down the track (a commission)? Would you knowingly allow the man with the backpack to wander onto the tracks on his own (an omission)? When psychologists turn these thought experiments into real experiments, they consistently find that people treat omissions as less bad than commissions. The key advantage to using a laboratory experiment is that the experimenter can make it so that there is no question as to whether an omission (like allowing the man with the backpack to wander onto the track) was 100 percent intended.

There are plenty of real-world examples, too. We're quick to berate someone for littering but hardly bat an eye when someone passes by litter without picking it up. We'd think it downright evil to steal from the homeless dude outside Starbucks, but we don't think it's evil to deny him the money we are about to use to buy ourselves that latte. In all such examples, failing to help is A-OK, while actively harming is absolutely not, even though the effect, and intent, are the same.

Why is our sense of ethics so unduly influenced by whether the harm was done via action or inaction? Why don't we simply attend to whether someone knew they could have done something helpful and then did or didn't do it?[4]

Indirect speech. In June 2019, the office of French president Emmanuel Macron posted a video of Ivanka Trump looking entirely out of her league at the G-20 summit of world leaders. Although the French of course denied it, the video was widely interpreted as an expert-level troll of the Trump administration and an example of the French's mastery of saying a whole lot without saying anything at all.[5]

Although we're not always as masterful as Macron, the rest of us are also prone to communicating indirectly. We raise an eyebrow to indicate skepticism and cough to attract attention. At the table, we don't tell Grandma to "pass the gravy," choosing, instead, the more roundabout: "If you could pass the gravy, that'd be awesome." We ask an employee, "Could I ask you to take this on?" when we're really just assigning them the task. We decline a romantic setup by saying that "Alex seems nice" or "has a great personality." We may not especially be keen on a drink when we politely ask, "Would you like to come up for a drink?" We communicate a threat using body language or by casually dropping that "it'd be a shame" if something bad happened. We offer bribes by asking questions like "Is there anything we can do to take care of this here?" We express our disapproval with the outcome of an argument by grunting, "Fine, whatever." Instead of declining a request, we feign to agree but fail to follow up. When mortal enemies like Israel and Iran negotiate, they don't meet directly but ask intermediaries to communicate on their behalf. *Etc. Etc. Etc.*

Such indirect communication comes at an obvious cost: it is more likely to result in miscommunication than the more direct options. "If you could pass the gravy, that'd be awesome" contains more words but not more information than "pass the gravy." "Alex seems nice" is usually understood as "I'm not attracted to Alex," but sometimes the implication is missed. Declining a request by feigning to agree can be downright misleading; travel guides to places where this practice is common (East Africa, for instance, and many parts of East and Southeast Asia) often feel the need to alert tourists and businesspeople so they are not angered by it. Intermediaries can, and often do, fail to accurately deliver a message or, as Shakespeare's Juliet learned the hard way, to deliver it at all. As they say, if you want something done right, do it yourself.

Why, then, do we choose to communicate indirectly via such hints or innuendos? Why are we passive-aggressive? Why don't we simply state our desires and intentions directly? Why do we "play telephone" with intermediaries and risk the message getting garbled or dropped, especially when a wrong message could lead to war?! Of course, the answer most people would give is that indirect communication is more polite or that it leaves room for saving face, but such proximate explanations, themselves, merit explanation. Why is indirect communication more polite? When is it important to communicate politely and, thus, indirectly? And what does it even mean to be polite or to save face?

THE CORE IDEA

The answer to all these puzzles boils down to just one insight. First-order beliefs—what you think happened—aren't all that matters for coordinated action. What you think others think (second-order beliefs) and what you think others think you think (third-order beliefs and so on) can matter as well.

We already hinted at how important such higher-order beliefs could be for coordination in Chapters 10 and 11. When playing the repeated prisoner's dilemma, if you saw your counterpart defect but thought he didn't notice, then you'd have been better off treating it as if he cooperated. In the norm enforcement game, if you saw a transgression but you alone were aware of it, then it wouldn't have made sense to punish the transgressor, since others would think you were punishing out of turn and you'd be punished. The logic can be applied to the hawk-dove game, too. If you thought you arrived first but thought your partner didn't realize this, then it would be better to play dove, lest you find yourself in a costly dispute.

In this chapter, we're going to explore this basic point using as our basic tool the state-signal coordination game combo introduced in the last chapter. Except, now, we will be modeling specific features of signals that influence higher-order beliefs to see how these features also influence coordinated action.

OBSERVABILITY

One feature of information that has an interesting effect on higher-order beliefs is observability. We already saw observability in action in Chapter 7, when we saw that subjects in laboratory experiments donated more if their donations were more likely to be seen by experimenters or fellow experiment participants.

So far, we only discussed the obvious effect observability has: it increases the chance you learn of a contribution or a transgression, which is essential if you are to react to them. However, observability also affects what you should do about a contribution or transgression *once you are aware of it*. If it wasn't so easy for you to learn of the transgression, then, even though you did learn about it, you'll think others haven't. You will have first-order confidence that the transgression occurred but will lack second-order confidence—confidence that others are confident that a transgression occurred. If sanctioning needs to be coordinated, then that matters as well.

Here's a simple game that illustrates this. As in the last chapter, the game has two players who represent observers deciding whether to sanction a norm violation, where this decision is modeled using the simple coordination game. As a reminder, in the coordination game players choose between two actions (sanction and don't sanction), and the payoffs are such that a player prefers to sanction if and only if the other is sanctioning with probability at least p.

Once again, we'll append a state-signal structure before this coordination game, which models what information players have when they make their punishment decision. All we need to do, then, is to define a state-signal structure that allows us to explore the influence of observability on second-order beliefs. Here's one that does that:

- There are two states, 0 and 1, where 1 indicates that the transgression occurred. The prior probability that the state is 1—the frequency with which transgressions occur—is μ.
- Each player independently gets a signal of either 0 or 1. When the state is 1, players get a 1 with probability $1 - \varepsilon$ and a 0 with probability ε. This means that when the transgression occurs, players usually learn about it but sometimes don't (and get a false negative). When the state is 0, players always get a 0 (there are no false positives).[6]

In this state-signal structure, $1 - \varepsilon$ can be interpreted as how observable the transgression is: when observability is 1, players always know whether the transgression occurred, and when it is 0, players always just see 0, so they can never learn anything about the transgression. Notice that when a player gets the signal of 1, regardless of ε, or how observable that signal was, they know for sure that a transgression occurred. But whether or not they also think the other player thinks a transgression occurred—their second-order beliefs—depends on ε.

The strategy we're interested in is the one in which players sanction only when they get a 1; that is, they sanction whenever they think a transgression occurred. Let's call this the conditional sanctioning strategy. Is it a Nash equilibrium for both to use the conditional sanctioning strategy? If all that mattered were first-order beliefs, the answer would be a definitive yes because, regardless of ε, when you get the signal of 1 you can be sure a transgression occurred. However, that's not what matters in this game. In fact, all that matters is how confident you are that the other will be sanctioning, which does depend on ε.

To see this, suppose that both players are playing the conditional sanctioning strategy. Can you benefit by deviating? Let's start by considering the case when you get the signal of 1. In this case, you will want to sanction if and only if you think the other is sanctioning with probability at least p. Since the other sanctions when they get a 1, this amounts to the odds the other got a 1 when you got a 1, which is just to $1 - \varepsilon$. So, as long as $1 - \varepsilon \geq p$, you won't want to deviate, which will be true so long as ε is small enough (observability is high enough).

What about when you get a signal of 0? The odds that the other will get a 0 as well, and therefore won't be sanctioning, is $1 - (\mu\varepsilon(1 - \varepsilon)) / (\mu\varepsilon + (1 - \mu))$, so you won't want to deviate, provided this is greater than $1 - p$. This condition will also hold if ε is small. So for conditional sanctioning to be Nash, we need transgressions to be easy to observe.

Thus, whether or not you can sanction a transgression *when you are aware of it* depends on how lucky you had to be to learn about it (how observable the transgression is, $1 - \varepsilon$).

Observability is the first feature of signals that impacts higher-order beliefs above and beyond first-order beliefs and, thus, impacts players' ability to condition on their signals. Let's now consider a second such feature of signals: how shared they are.

SHARED SIGNALS

In the previous model, each person's signal was determined independently (when the state was 1, the probability that each player saw a one was $1 - \varepsilon$, regardless of what the other saw). However, one could imagine instances where players' signals are not independently determined. For instance, the signal they observe might be part of a shared experience, publicly displayed, or easy to relay. In those cases, it will be pretty easy to condition on this signal. That's true even if the signal was unlikely to have been generated in the first place because of low observability. Maybe the signal is a photo that will reveal the transgression only if someone happens to take it at just the right moment, or a record that is kept secret and will be sent only if someone happens to forward the wrong thing to the wrong person. Although such signals are unlikely to be sent, once they have been, they can be shared far and wide.

We can see this once again with a fairly simple model. Actually, we will just need to tweak our previous model to allow signals to be correlated. Here's what it looks like:

- States are determined and interpreted as before (there are two states, 0 and 1, and the prior probability that the state is 1 is μ).
- Each player gets a signal of either 0 or 1. When the state is 1, players get a 1 with probability $1 - \varepsilon$ and a 0 with probability ε. These errors, however, are not necessarily independent. Their correlation is summarized by a parameter ρ (rho) that varies from 0 to 1. 0 means no correlation— that's the case we already considered, in which signals are independent. State 1 means signals are perfectly correlated—players always see the same exact signal.

The conditional sanctioning strategy (sanction when you get a 1) is now an equilibrium whenever $(1 - \varepsilon)$ or ρ is high. Technically, whenever $1 - \varepsilon (1 - \rho) \geq p$. That is, once signals are sufficiently likely to be shared, it's possible to condition on them, and that's true even if, due to low observability, a signal of 1 is unlikely to have been sent in the first place.

Notice, once again, that ρ influences higher-order beliefs above and beyond first-order beliefs: whenever you observe the signal of 1, you are sure a transgression occurs, but whether or not you think others are also sure depends on how correlated your signals are (as well as how observable the signal was).

We now have seen two distinct ways signals can differ in terms of higher-order beliefs and the influence this has on whether the signals can be utilized for coordinated action. Let's take a look at a third.

PLAUSIBLE DENIABILITY

Let's start by illustrating what we mean by plausible deniability with a classic social psychology experiment from the 1970s. In the experiment, subjects were brought to the lab ostensibly to provide their opinion on some rather old silent films. When they arrived, they were asked to choose a seat in front of one of two TVs. In front of each TV there were two seats: one of the four was occupied by another subject (actually a confederate hired to play along) who wore a metal brace designed to make him seem handicapped.

When the two TVs played the same movie, three-quarters of subjects chose the seat next to the handicapped confederate, lest they appear bigoted. But when the TVs showed different movies, the results flipped: now three-quarters of subjects chose not to sit next to the handicapped confederate.[7] The interpretation usually given is that in the case where the TVs are playing the same movie, there is no plausible deniability: the only explanation for choosing not to sit next to the handicapped confederate is, uh, bigotry. If, however, the TVs are playing different movies, this provides a benign explanation for choosing to avoid a handicapped person— maybe the subject just prefers the other film. (Mind you, it wasn't much of an exculpatory explanation. The experimenters chose very similar movies. Both were silent films. About clowns. And, of course, the experimenters were careful to ensure their result held up regardless of which movie the handicapped confederate was seated in front of.)

The study illustrates a common phenomenon: the very existence of an exculpatory explanation like "I prefer the film playing on the other TV" can, regardless of how unlikely it is, serve as cover for making a decision that would otherwise have been penalized. Why would that be?

Here's the intuition. Let's suppose that we do sanction bigots whenever we are sufficiently confident they are bigoted, even if there is some exculpatory explanation. How confident? Let's say 95 percent. Now let's consider what would happen when you are just barely 95 percent confident that the exculpatory explanation isn't valid and the person is in fact a bigot. What made you 95 percent confident? Maybe you know something about the two films and know that it's unlikely that someone would have a strict preference for one over the other. Maybe you have some baseline suspicions that this person is a bigot. OK. That requires some knowledge about films and bigots. Do others have this knowledge? Maybe yes, maybe no. But—and here's the kicker—we are considering a case where you are 95 percent certain. That's pretty darn certain. That requires a pretty hefty degree of private knowledge, and you have to be pretty lucky to have acquired all that private knowledge. Odds are, others will be less knowledgeable than you in that case. So you will suspect they are less confident than you are. That's enough to make you tempted to deviate, and enough to prove this is not an equilibrium.

In contrast, when there is no exculpatory explanation, say the two rooms were playing the same film, you don't need any private information to be confident the person is a bigot. And so you will be sufficiently confident others are confident the person is a bigot. There is no problem sanctioning in these cases.

Now let's formalize this. Once again, we'll model the sanctioning decision of two observers using a coordination game and tack on a state-signal structure to describe the information available to them when deciding whether to sanction. This state-signal structure will work a bit differently than the last two. We'll present it to you first and then explain why we're using it:

- States are determined and interpreted as before (there are two states, 0 and 1, and the prior probability that the state is 1 is μ).
- Each player gets two signals. The first is a fully shared signal and takes on a value of 0 or 1. When the state is 0, the players see 0 with probability $1 - \varepsilon$ and 1 with probability ε. (This means that the players sometimes see a false positive.) When the state is 1, players always see 1 (there are no false negatives).

- However, players don't know ε. They know the distribution of ε, f. They also each get an independent signal of ε, ε_i, whose distribution we'll call g_ε. For instance, f might be the uniform distribution between 0 and 1. g_ε might be the triangle distribution with peak at ε and endpoints at 0 and 1, meaning players' signals can take any value between 0 and 1 but become less likely (at a constant rate) as they are further from the true ε.

As you can see, we've switched to focusing on false positives. This is because they naturally relate to exculpatory explanations: whenever players see a 1, they know there's some chance that the exculpatory explanation is true and the state is 0.

However, players don't actually know exactly how likely the exculpatory explanation is to be the true explanation. They have a guess—which can be a very good guess but is a guess nonetheless. ε_i represents that guess.

The question is: Can players sanction only when that guess leads them to believe the exculpatory explanation is sufficiently unlikely? That is, can they sanction if and only if ε_i is below some threshold, say, $\bar{\varepsilon}$? This amounts to sanctioning whenever they are sufficiently confident that the state is 1.

As we already hinted, the answer is, generally, no. We can see this by focusing on the case where $p = .5$. Whenever a player gets a signal ε_i just below $\bar{\varepsilon}$, they would need to be 50 percent confident that the other player did as well. However, this will only be true when $\bar{\varepsilon}$ is exactly .5. If any lower threshold were chosen—if we required the players to be more confident than 50 percent to sanction—they would be less than 50 percent confident the other got a signal below this threshold and so would have an incentive to deviate. Had we chosen p less than .5, the equilibrium $\bar{\varepsilon}$ would be below .5 but still couldn't be too extreme—it would still not be possible to sanction only when one is *very* confident. This result depended on the fact that, given f and g_ε, players' signals exhibit regression toward the mean, which causes a player who gets extreme signals to assume the other player is unlikely to get equally extreme signals.

So, with that, we see that being highly certain but not absolutely certain is a third way that first-order beliefs can be high while higher-order beliefs are not quite as high, and this will preclude coordinated action.

HIGHER-ORDER UNCERTAINTY

So far, we've focused on identifying ways that you might know things but not be sufficiently confident others do, too—second-order beliefs. Sometimes, though, you get cues as to what others know. Does this solve the problem? Not necessarily. Not if those cues are also not terribly observable, shared, or undeniable.

To see why, we'll focus on the case where those cues aren't so observable. The intuition is: even if you know the other knows, the other won't be confident you know she knows. So she will act as if you don't. Let's now explore this more formally.

The game again has two players whose decision of whether to sanction is represented by the coordination game.

- There are again two states, 0 and 1, which represent whether the transgression occurred.
- Player 2's signals work as before: when the state is 1, she gets the signal of 1 with probability $1 - \varepsilon$, and 0 otherwise. When the state is 0, she always gets a signal of 0.
- As for player 1, we assume he always learns the true state (for simplicity) but also gets a signal as to player 2's signal: whenever player 2 gets the signal of 1, player 1 will observe a signal of 1 with probability $1 - \delta$, and 0 otherwise. Whenever player 2 gets a 0, player 1 does, too.

Is there an equilibrium for the two players to sanction when they each get a signal of 1? In that case, both players would know the state—first-order knowledge—as well as the fact that the other player knows the state—second-order knowledge. Is this sufficient?

The answer depends on δ (and no longer just on ε). Only if $1 - \delta$ is high enough (greater than p), will player 2 be sufficiently assured that player 1 is aware that she got the signal and, so, will be punishing.

Here we've seen that third-order beliefs (what player 2 thinks player 1 thinks she thinks) also affect coordinated action. Above and beyond second- and first-order beliefs. This same logic would extend to any order of uncertainty. If the other player gets a cue that you got a cue about his signal of the state, that helps only if that cue is observable. In short, higher-order uncertainty matters, too.

MAPPING TO OUR PUZZLES

So far we've seen four ways higher-order beliefs come into play and influence players' ability to coordinate:

- **Observability.** How easy it was to learn of a transgression?
- **Correlation.** Do others have access to the same sources as you did?
- **Plausible deniability.** Are you fairly but not absolutely certain the exculpatory explanation isn't valid?
- **Higher-order uncertainty.** Even if you know, and you know what others know, do they know that you know?

When you learn something that wasn't easy to observe, you assume you chanced upon it and others won't be so lucky. When you learn something via a private channel, you cannot rely on the fact others will be privy to the same info. When you are fairly certain but not absolutely certain, odds are others are less certain than you are. And when you know what others know, but they think it's unlikely that you know what they know, they will still act as if you don't know what they know.

Our next task is to see what role higher-order beliefs and co-ordinated action play in our motivating puzzles. We will see, for instance, that symbolic gestures take advantage of highly observable, correlated, and undeniable signals, in order to enable people to switch to a new equilibrium. Intentions, as compared to actions, are much harder to know, more likely to become known via private channels, and more deniable. This makes transgressions of omission much harder to sanction than corresponding commissions. Innuendos take advantage of higher-order uncertainty, as well as plausible deniability, to preclude the possibility of switching to a less desirable equilibrium if things don't go as planned.

SYMBOLIC GESTURES

Symbolic gestures are puzzling because things that don't matter *matter*: we put enormous weight on mere words like *I'm sorry* and *I love you* and spare no expense on rituals, ceremonies, and elaborate displays that don't convey any new information.

Our explanation for this puzzle is that things that might otherwise not matter can matter when coordination is important and when those things are observable, public, and undeniable. And that symbolic gestures satisfy these criteria.

We've already seen an example of a symbolic gesture at work when we discussed apologies in Chapter 5. An apology fits our explanation relatively neatly. In the hawk-dove game and the real-life settings it is intended to describe, coordination is important: it's important not to play hawk when you're expected to play dove and vice versa. So, mere words that would otherwise not matter can help players coordinate a switch from (hawk, dove) to (dove, hawk). This explanation isn't different from the one we gave in Chapter 5, where we talked about uncorrelated asymmetries establishing expectations—it's just stated a bit differently.

We are, however, adding that, to have the desired effect, apologies should have certain features, namely, they should be observable, are best made publicly, and must be undeniable. Consistent with this, we see that people often insist on an explicit apology ("Say that you're sorry!") and sometimes even that this apology be made in front of others ("Don't tell me, tell the whole class!"). They also scrutinize the language used in the apology, insisting that the apologizer not just express regret or admit to having made a mistake, but that they also take responsibility ("Say it was your fault!") and commit to changing their behavior ("Say that you won't do it again!"), and often, even, that they use the words "I'm sorry" or "I apologize" when doing so.

As long as apologies have our three features, that's all that's needed. It doesn't matter if they're coerced or given freely. Nor does it matter if they're heartfelt or given through gritted teeth. It also doesn't matter if they contain no additional information other than "I'm sorry." All that matters for them to precipitate coordinated action is that they are made explicitly and said loud and clear so that everyone can hear.

When apologies don't have our three features, they're treated with skepticism, presumably because it is unclear whether the offending party will actually mend their ways and, also, therefore, whether it's optimal to act as if they will. Hillary Clinton's apology to Pakistan offers a particularly interesting case of such a nonapology. At the onset of the crisis, Pakistan insisted upon a real apology, presumably with the intention of switching who could play hawk over Pakistani airspace. However, eventually, Pakistan backed down from this request, insisting only that the United States make some kind of public statement involving the word *sorry*, which we presume was an effort to save face. A Pakistani team worked together with Clinton's team to craft her statement, which read:

> We are sorry for the losses of the Pakistani military. We are committed to working with Pakistan and Afghanistan to prevent this from ever happening again.[8]

This carefully worded nonapology did not take responsibility for the Pakistani deaths or commit the US to changing its behavior:

"we are sorry for the losses" avoids taking the blame for those losses; "committed to working with Pakistan and Afghanistan" is not the same as committing to prevent such a thing from ever happening again. This did not escape the attention of critics in Pakistan, but their voices were drowned out by the billions in aid that came along with Mrs. Clinton's statement.

Let's move on from apologies and explore how other types of performative speech—words that do things[9]—are used to change the nature of a relationship.

Alan Fiske has famously suggested that people tend to view relationships as falling into one of four bins. Here are Fiske's four bins, in his own words.[10]

- **Communal sharing (CS)** relationships are based on a conception of some bounded group of people as equivalent and undifferentiated. In this kind of relationship, the members of a group or dyad treat each other as all the same, focusing on commonalities and disregarding distinct individual identities. People in a CS relationship often think of themselves as sharing some common substance (e.g., "blood"), and hence think that it is natural to be relatively kind and altruistic to people of their own kind. Close kinship ties usually involve a major CS component, as does intense love; ethnic and national identities and even minimal groups are more attenuated forms of CS.

- **Authority ranking (AR)** relationships are based on a model of asymmetry among people who are linearly ordered along some hierarchical social dimension. The salient social fact in an AR relationship is whether a person is above or below each other person. People higher in rank have prestige, prerogatives, and privileges that their inferiors lack, but subordinates are often entitled to protection and pastoral care.

- **Equality matching (EM)** relationships are based on a model of even balance and one-for-one correspondence, as in turn taking, egalitarian distributive justice, in-kind reciprocity, tit-for-tat retaliation, eye-for-an-eye revenge, or compensation by equal replacement. People are primarily concerned about whether an EM relationship is balanced and keep track of how far out of balance it is.

- **Market pricing (MP)** relationships are based on a model of proportionality in social relationships; people attend to ratios and rates. People in an MP relationship usually reduce all the relevant features and components under consideration to a single value or utility metric that allows the comparison of many qualitatively and quantitatively diverse factors.

Relationship bins like these require coordination. It won't work out so well if one party treats the relationship as authority ranking, and the other starts acting like it's one of equality matching. So how do you change the terms of the relationship, without risking such costly miscoordination?

That's the kind of thing mere words can help with. Explicit statements like "I love you," "I do," "You're fired," "We are no longer friends," or "I am breaking up with you" can drastically alter the terms of a relationship. Importantly though, they alter the terms of the relationship in a way that avoids miscoordination: we both know what those words mean and what kind of relationship we now do or don't have. Uttering them minimizes the risk of one of us continuing to treat the relationship as if nothing has changed.

One thing that's crucial, though, is for these pronouncements to be undeniable—for there to be no ambiguity. And if these pronouncements have one thing in common, it's their lack of ambiguity. Once your boss has uttered, "You're fired," you can't just claim you thought he meant you should take a breather. That's why "I am breaking up with you" can be so painful and why "But you know how I feel about you" isn't the desired response to "Do you love me?"

As with mere words, physical gestures—handshakes, bows, curtsies, a first kiss, holding hands, or stomping on a glass under the *chuppah*—can alter the terms of a relationship. Such gestures are, by their very nature, shared and undeniable experiences. A kiss is not something one person can easily brush off and pretend didn't happen. Nor is it something that happens by accident. Same with holding hands. It's not like one person's hand could just slip into another's and stay there. There isn't really an exculpatory explanation. Moreover, often these romantic gestures are conspicuously displayed, as those forced to observe PDAs (public displays of affection) will readily attest. PDAs are a good way to make one's relationship clear to onlookers. Likewise, bows or curtsies are hard to deny once having been performed, are easily recognized by anyone in the room, and thus are good ways to make commonly known who is submitting to whom. Honorifics and titles, like *sir*, *madam*, *professor*, *doctor*, or *Mr. President*, or those appended to a name, like *san* (Japanese) and *ji* (Hindi), can help with this, too. That's likely why we pay so much attention to them. In some languages, we immediately and constantly label relationship types by carefully applying formal and informal pronouns like *vous* and *tu*. In some cultures, we do the same by calling elders auntie and uncle, while calling equals *brother*, *sister*, *cousin*, or just *bruh*, *sis*, and *cuz*.

Elaborate public rituals often serve a similar purpose, although they tend to be about coordination among more players. We didn't explicitly model many player games, but the same insights should hold: only when signals are highly observable, public, and undeniable can they permit large scale cooperation. Ritualized public ceremonies do the trick. For instance, coronations make clear that everyone will treat the same person as their ruler, follow the ruler's orders, and punish those who disobey. An event like a coronation is typically open to the public; indeed, efforts are made to ensure that as many spectators as possible can attend. The events are also filled with easily recognizable gestures (like anointment) or items that have been canonized over time (like the Stone of Scone). These help to make the event undeniable and ensure that the enthroned is the undisputed ruler.

Marriage ceremonies aren't much different, though—Bollywood movie weddings excepted—they're usually on a smaller scale. Like coronations, weddings have exaggerated and undeniable ritual elements that are easy for everyone to see (walking down the aisle, standing under a *chuppah*) and hear ("Do you take this . . ." "You may now kiss the bride"). They also typically include many members from the community. Of course, weddings aren't meant to help ensure that we all obey the same leader. Rather, they are meant to identify who is no longer available as a potential romantic partner and, in some societies, who is obligated to support or obey whom or which clans are allied with which. As with apologies, weddings can do these things even if the words pronounced aren't genuinely felt. (Until death do us part?! With today's divorce rates? And isn't this your third marriage?)

Of course, coronations, weddings, and many other ritualized ceremonies have elements of costly signaling at play. Once you've invited everyone in the community over, you might as well also take advantage of the occasion to show off how many supporters you have or how much money you can afford to throw away. But that's a separate phenomenon. It explains the mountains of flowers and shrimp scampi, not the "I dos" and other highly ritualized elements of the ceremonies. The thing that brings everyone to the same place in the first place is the need to affect coordinated action. The scampi is just a bonus.

OMISSION-COMMISSION DISTINCTION

Next let's see how our models help us to understand some quirks of morality and altruism, starting with the omission-commission distinction. Think back to the story of the *Struma*. As a reminder, the puzzle here is that we care not just about whether someone intentionally transgressed (caused hundreds of Jews to perish) but also whether they did so via action (by gassing them) or inaction (refusing to allow them off the boat before it was towed to sea). Why?[11]

Imagine trying to eliminate the omission-commission distinction. This would require treating intentional omissions (punishing the English and the Turks for refusing to allow the Jews off the boat, knowing that the Russians would torpedo it) but not unintentional inaction (not punishing when a boat gets torpedoed and no one had the ability to prevent it). To punish intentional omissions, we would therefore need to be able to condition on intent.

Therein lies the problem. Intent is typically hard to observe, never really public, and almost always deniable.

To observe intent, we have to have just the right information. Who made the decision to tow the boat to sea? Did they know that unmarked boats are likely targets for Russian submarines? This information is often only discovered in seclusion—top-secret meetings, KGB archives, whispers shared among military wives who overhear their husbands' jabber. And we must guess at the Brits' and Turks' intent with the help of hard-to-share, hard-to-observe, somewhat deniable private information: we might know from experience how reliable our source is, the source may mention discussions among British officials acknowledging the risk to the ship, and so on.

The same holds true for runaway trolleys. If we see someone omit pulling the lever, we might happen to know he actively chose not to, maybe because we know he hesitated, looked back and forth, thought twice, and then sat back down. But would others be so sure he made an active choice to let the trolley barrel on? Maybe he just wasn't paying attention? Maybe he tried to pull the lever but couldn't get to it in time?

Actions, on the other hand, are observable, easily made public, and not really deniable, at least in comparison to intentions. The English and the Turks could have executed the Jews the way the Nazis did. While it's likely this would be done in secret, it would eventually have come to light, as happened when photos from Auschwitz were leaked and mass graves in the Ukraine were discovered. Such information could be made widely accessible and would turn into a public signal. Moreover, there wouldn't be any thorny matters of interpretation. It'd just be a matter of figuring out who pulled the trigger and who gave the order. There would be no real question of who knew what and what motivated them. Similarly with the lever in the trolley study. If it's pulled, it's pulled. If that's on camera, it can be made public. Or someone can check the logs to see how the tracks were aligned at the start of the shift. People don't pull levers by accident just at the moment a trolley is barreling down the track. Because of this, unlike intentions, we can condition on actions. That is, we can punish commissions.

This analysis leads us to a couple of predictions that seem to bear out in practice.

The first: if we can't condition on intentions, that not only means we can't punish intended omissions, but it also means that we have to punish unintended commissions. The real-life story of Donnie Brasco and Sonny Napolitano offers a gruesome confirmation of this prediction. Brasco was an undercover FBI agent who infiltrated a New York crime family by befriending Sonny Napolitano and members of his crew. After Brasco's identity was revealed, Napolitano's boss ordered Napolitano killed and had his hands chopped off, even though Napolitano was a loyal soldier who, obviously, did not intend to befriend an FBI agent![12] Sadly for Napolitano, his intentions, though obvious, were still private and not undeniable.

The second prediction: intent will be taken into account when coordination plays less of a role. Before we look for evidence for this prediction, let's briefly discuss the role that coordination plays in some of our examples. When it comes to protecting human rights—or failing to, as in the case of the *Struma*—coordination plays a role because there is no single body that can unilaterally enforce these rights; they must, instead, be enforced via international sanctions, war, or public outrage, which must be coordinated among the world powers or the people on the streets. The United Nations International Court of Justice at The Hague was intended to solve this issue but, in practice, has no way to enforce its judgments when a party does not voluntarily comply. In Sonny

Napolitano's case, his capos couldn't exactly depend on Uncle Sam to enforce their codes. They were dependent on informal, coordinated enforcement by those inside the organization. And the psychology that gets tapped into in trolley problems arguably stems from moral intuitions that get socially enforced, à la our norm enforcement model.

Now let's consider some instances where coordination plays less of a role. For instance, imagine witnessing someone fail to help a friend of hers, say, by failing to respond to a text message asking for help moving to a new apartment until it's too late to be relevant, and then playing it as if she didn't see the message in time but would've been happy to help ("Hey, really sorry. Do you still need a hand?"). Would you speak out and say something about this? Perhaps, but maybe not, because she has plausible deniability and speaking out is a form of punishment. Either way, though, you'd probably think to yourself that so-and-so isn't such a great friend, and you trust her a bit less to come through if you yourself need a favor sometime. That's because deciding whether to trust so-and-so isn't really about coordination. Trust is a personal assessment, one that you can make on your own. For such personal assessments—and the decisions that build off these, like whether to rely on this friend in the future—it doesn't matter that intentions aren't observable, public, or undeniable. What matters is what you know. If you have decent information about intentions, you can use it.

Our dealings with kin provide another instance where coordination plays less of a role. As discussed in the last chapter, our love for our kin is driven by kin selection—we evolved to care about our kin because we share so many genes with them. Higher-order beliefs and coordination don't matter in this case. What matters is whether our kids or cousins are hungry or otherwise need help, not what other people think or whether we act the same way as they would. Whether our kin fail to obtain our help by omission or commission is immaterial. So, just as we saw in the last chapter, the comparison between altruism toward strangers and altruism toward kin is particularly telling.

Robert Kurzban, Peter DeScioli, and Daniel Fein have in fact used trolley problems to show that while people typically say they wouldn't push a stranger with a heavy backpack in front of the trolley to save five strangers, they would push a brother with a heavy backpack in front of the trolley to save five brothers. When it comes to brothers, it doesn't matter that a harmful commission is required, just the number of brothers who will survive the disaster.[13]

We similarly expect the omission-commission distinction to diminish in settings where a single authority can enforce the law unilaterally. Unlike international bodies tasked with preventing genocides like the *Struma*, domestic authorities can focus on the amount of harm done, consider intentions (as they do, for instance, when distinguishing murder from manslaughter), and ignore whether the harm was done by omission or commission.

◆

BEFORE WE MOVE ON TO ADDRESSING INDIRECT COMMUNICATION, WE will use our analyses to tackle a few bonus puzzles that are similar in flavor to the omission-commission distinction.

Avoiding the ask. Each winter, as the holidays approach, Salvation Army volunteers can be found outside heavily trafficked retail establishments. They wear red Santa hats and collect donations in a red bucket, ringing a bell to get attention from passers-by. In a clever experiment, three behavioral economists teamed up with some Salvation Army volunteers to test strategies for increasing the donations they received.[14] The researchers wondered if the Salvation Army volunteers might increase donations by explicitly asking, "Can you spare some change for the Salvation Army today?" So the volunteers were trained to sometimes ask and sometimes not and assigned to do so according to a schedule that was randomized to allow the researchers to determine the effect of the explicit ask.

You might think that asking someone a question like "Can you spare some change for the Salvation Army?" wouldn't matter. After all, the question contains no new information about the Salvation Army, which is a well-known organization. Yet, when the volunteers employed their explicit asks, more passers-by donated, and their red buckets filled more quickly. This seemed promising, but before the researchers could rest on their laurels, they noticed a problem: people were avoiding the Salvation Army volunteer at the front door by going in and out of the store through the side door—a small, utility door normally used to access the recycling bins. But don't judge! If, as you've walked down the sidewalk, you've pulled out your phone so you can feign being busy instead of donating to a panhandler then you, too, are guilty of avoiding the ask.

Jason Dana, Daylian Cain, and Robyn Dawes developed a clever experimental paradigm that can be used to cleanly demonstrate that people who otherwise would have given will readily avoid the ask.[15] In their experiment, subjects play a simple dictator game—remember, that's the one in which you're given a few bucks and asked how much you'd like to give to someone else. In their experiment, Dana and Cain followed this dictator game with a surprise. They offered subjects the option of paying a dollar so that their partner would never find out the game had been played. Not only did a large fraction of subjects—about a third—take them up on this offer, but many of those subjects had also already agreed to share a meaningful chunk of their winnings with their partners.

So why do we give when asked but go out of our way to avoid being asked?

Strategic ignorance. One problem that public health workers around the world have encountered when combating HIV is that people don't get tested for sexually transmitted diseases like HIV often enough. The problem is particularly acute among individuals who engage in high-risk activities, for whom frequent testing is particularly beneficial. It's sometimes thought that these individuals prefer to remain strategically ignorant since, if they do end up having HIV, they know they will feel obliged to change their ways.

But why, then, don't they feel equally obliged to find out if they have HIV?

One of the researchers we just mentioned, Jason Dana, conducted another great experiment, this time together with Roberto Weber and Jason Kuang, in which they show strategic ignorance in the lab.[16] Subjects in their experiment were given a pot of money and asked to choose between two possibilities. Option A split the funds between them and another subject evenly—five dollars each. Option B gave the first subject a bit more money (six dollars instead of five), but reduced the other subject's take all the way to one dollar. Three-quarters chose the selfless option and split the funds evenly. For subjects in another treatment, everything was exactly the same except they couldn't tell whether the self-interested option was also antisocial because the other's payout was obscured. Before making their choice, subjects could (costlessly!) uncover the other's payout or not. About half didn't bother. They simply chose the option that paid them more.

The subjects in the experiment behaved just like individuals who are at high risk of HIV but do not get tested: the same individuals who would feel obligated to avoid knowingly behaving antisocially choose not to know.

Means–by-product distinction. One major complaint that Israel levies against its nemesis Hamas, the Islamist organization that controls the Gaza Strip, is that Hamas uses human shields to protect military assets by hiding its military storehouses in dense residential areas, launching attacks from these areas, and then, when Israel warns residents of upcoming reprisals, encouraging residents to ignore these warnings. Indeed, Hamas admits to much of this, which, in most people's eyes, makes them rather villainous. Indeed, the use of human shields is a violation of the 1864 Geneva Conventions and is considered a war crime by the United Nations. Bad stuff.

At the same time, Israel—and, really, any party engaged in war—does sometimes attack targets even when it knows there's a meaningful risk of civilian casualties. This is bad, but most people feel it's less bad than using human shields. And it's not a violation of the Geneva Conventions nor is it a war crime.

The weird thing about this is that in both cases, civilians are being knowingly put in harm's way. The only difference is that in one, that harm is a means to an end (preventing an attack), whereas in another it's a by-product (civilians aren't explicitly the target, but their presence isn't important enough to prevent the attack).

Why would we distinguish between knowingly causing harm as a means versus as a by-product? Again, why don't we simply attend to the amount of harm that was knowingly done?

This distinction between means and by-products also appears in philosophy and interpretations of religious scripture. The thirteenth-century Christian philosopher Thomas Aquinas, taught in every liberal college, argued that it was acceptable to, for instance, kill a man in self-defense (as a by-product) but not as a means to saving one's own life. Likewise, the Talmud, which predates Aquinas by some eight hundred years, asserts that "if two Jews are stranded in the desert, with only enough water for one of them to survive, and one is currently holding that bottle of water, Rabbi Akiva's opinion is that the 'water carrier' should drink the water himself, and not give the water to his friend because 'your life takes precedence.'"[17] It is not, however, permitted to kill the other person to take possession of the bottle of water. Killing him as a by-product of drinking the water? Yea. Killing him as a means to drinking it? Nay.

There are also trolley experiments that neatly demonstrate this moral quirk. In one variation, by Mark Hauser, Fiery Cushman, Liane Young, R. Kang-Xing Jin, and John Mikhail, a bystander can prevent the out-of-control trolley from hitting five people by sending it down a side track, where it will be stopped by a heavy weight before returning to the main track where the five people lie. In the first treatment, there is a man standing in front of the weight who will, tragically, be killed as a by-product. In the second treatment, the heavy weight is the man himself, who would now be killed as a means. Thirty percent more people think it's permissible to change the tracks in the first treatment than the second.[18]

Same outcome, equally foreseeable in both cases, yet people feel a moral distinction. Why?

◆

OUR EXPLANATION FOR THESE MORAL QUIRKS IS SIMILAR TO THE EX-planation we gave for the classic 1970s study with the handicapped confederates. As a reminder, in that study, the very existence of a benign explanation like "I prefer the film playing on the other TV" could, regardless of how unlikely it is, serve as cover for making a decision that would otherwise have been penalized. The reason we gave for this was that the bigoted behavior is deniable in the separate film treatment: you might be quite confident those who choose the other film are bigoted, but how confident you are depends on all sorts of details and priors, like your sense of how different the films are and how likely it is someone would prefer one film to the next. Since your beliefs are rather extreme, your expectation is that others won't be quite as confident as you (there's deniability). Moreover, these details and priors are difficult to come by (low observability) and aren't shared or easy to share.

Our explanations for the remainder of the moral quirks are not much different. Think about the kind of information you'd need to rule out benign explanations for going out the supermarket door that's far from the Salvation Army volunteer, for failing to show up at the STD clinic, or for bombing a building that happened to also have civilians inside. Now think about the information you'd need to rule out benign explanations for saying no to the Salvation Army volunteer after an explicit ask, for engaging in unprotected sex after testing positive for HIV, or for forcing a group of civilians into a building that you knew was going to be bombed. In which cases is the information needed to rule out the benign explanation observable, easily sharable, and undeniable?

INDIRECT COMMUNICATION

The last of our common knowledge applications is indirect communication: why we use innuendo and intermediaries, and why we are so often polite or passive-aggressive. To explain these, we will take advantage of the fact that higher-order beliefs matter for coordinated action, while what people know—mere first-order beliefs—matters for other settings. What happens when you want to influence behavior in the latter but not risk altering behavior in the former? You want to send a signal that utilizes one of the key features we have modeled, like plausible deniability or higher-order uncertainty. That will allow you to alter first-order beliefs without influencing higher-order beliefs, and will allow you to have your cake and eat it too.

When asking someone, "If you could pass the gravy, that'd be awesome," "Would you like to come up for a drink?" or "Is there anything we can do to take care of this here?" we typically are motivated to communicate *something*. If the other person knows we want the gravy, a romantic relationship, or to offer a bribe, he or she will be much more likely to pass the gravy, come upstairs, or accept the bribe. Similarly for a hostage negotiation via an intermediary, we want the hostage taker to know we're willing to exchange funds for those hostages. In these examples, there is information we want to convey for the purposes of influencing behavior that has no coordination element.

But at the same time, there is a risk. We want grandma to pass the gravy, not to think that we think we can boss her around. We want Alex to come up if he is also interested in a romantic relationship, but we also want to save face if Alex declines—perhaps even to continue to hang out with Alex as a friend or coworker, as if nothing different was ever on offer, or to avoid social sanctioning, if the ask was illicit. We want the cop to accept our bribe if he's willing but to minimize the chances of sanctioning if, by chance, the conversation is overheard or the cop is on the level. We want to try to bring the captives home but to avoid setting a new bargaining equilibrium where we are no longer expected to stand firm at "we don't negotiate with terrorists" when it comes to future rogue states or terrorists.

Indirect speech addresses the speaker's dueling wants because it communicates information but does so in a way that precludes higher-order beliefs. Did John attempt to bribe the cop? The cop might know the answer because he could see the bills protruding from John's wallet. But could a second cop on the scene see the bills (is it observable)? And did that cop manage to get it on camera (is it shared or shareable)? And was it protruding enough that the cop was *sure* that John had pulled it out on purpose (is it undeniable)? Or what about that invitation to come up for a drink? Is Alex pretty sure his friend is asking for a romantic relationship and so can respond accordingly, if he is interested in the same? But how does he know that is what the other is asking? Body language, earlier hints in the conversation, tone of voice? All these cues are hard to observe, not seen equally by everyone at all vantage points, hard to relate to third parties, and quite deniable. Good enough for Alex to decide whether to take his friend up on the offer. Not good enough to alter the relationship or get himself into trouble if he's not.

And that's the point: by speaking indirectly we are able to affect first-order beliefs without bringing higher-order beliefs along for the ride, thereby enabling us to affect the other's behavior when coordination is not involved without risking any effect on coordinated behavior. Speaking indirectly allows us to have our cake and eat it too.

THESE ARE JUST SOME APPLICATIONS THAT CAN BE EXPLAINED BY THE role of higher-order beliefs in enabling or preventing coordinated action. And just some aspects of signals that affect higher-order beliefs and, thus, our ability to use signals when coordinating. Hopefully, though, by this point, our message is clear: higher-order beliefs matter when coordination plays a role.

Next up: Where does our sense of justice come from and why is it so weird? The answer lies with subgame perfection.

How observable and shared are signals?

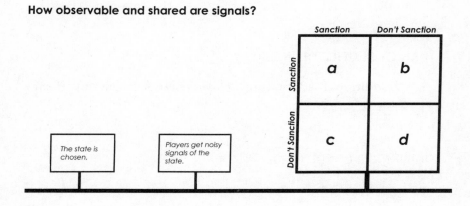

SETUP:

- First, the state of the world is chosen. It is 1 with some probability μ and 0 otherwise. The state could indicate whether a transgression occurred, with 1 indicating that it did.
- Then, each player receives a noisy signal of the state, with the following properties:
  - No false positives: when the state is 0, players receive signal 0 with certainty.
  - False negatives: when the state is 1, the signal is equal to 1 with probability $1 - \varepsilon$. ε represents the rate of false negatives.
  - Players' signals can be correlated. The correlation is indicated by ρ. $\rho = 0$ indicates the signals are generated independently. $\rho = 1$ indicates the signals are always identical.

(Technically we define ρ as $\rho = (r (1 - \varepsilon) - (1 - \varepsilon)^2) / (1 - \varepsilon - (1 - \varepsilon)^2)$, where r is the probability that player 1 obtains a signal of 1 given that the other player has received a signal of 1. This follows the standard definition of conditional correlation.)

- Finally, players play a coordination game. The parameter $p = (d - b) / (a - c + d - b)$ indicates how confident players must be that the other is playing A before they would want to play A.

STRATEGY PROFILE OF INTEREST:

- "Conditional sanctioning": players play A if and only if they see a 1.

EQUILIBRIUM CONDITIONS:

- $1 - \varepsilon(1 - \rho) \geq p$.
- $(\mu\varepsilon(1 - \varepsilon)) / (\mu\varepsilon + (1 - \mu)) \leq p$. This condition will not bind so long as the likelihood of a transgression, μ, is relatively small.

INTERPRETATION:

- It is possible to coordinate on the basis of a signal if the signal is observable (low ε) or shared (higher ρ).

Plausible Deniability

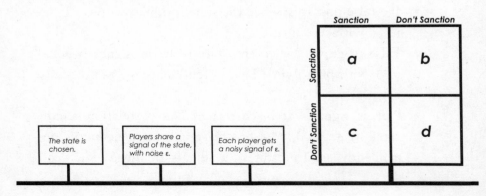

SETUP:

- First the state is chosen. It is 1 with some probability μ and 0 otherwise.
- Players get a signal of the state, which can be either 0 or 1, again. They also get a signal of the rate of false positives.
  - The signal of the state is fully shared.
  - The signal of the state has no false negatives: when the state is 1, players receive signal 1 with certainty.
  - The signal of the state has false positives with rate ε; when the state is 0, the signal is equal to 1 with probability ε.
  - Players don't directly observe ε. Instead, they know ε is distributed according to some known distribution f. And they each observe a signal of ε, ε_i, which itself is distributed according to some distribution, g_ε, which may depend on ε.

STRATEGY PROFILE OF INTEREST:

- Players play A if and only if they receive 1 and observe $\bar{\varepsilon}_i > \bar{\varepsilon}$ for some $\bar{\varepsilon}$.

EQUILIBRIUM CONDITION:

- Provided the distributions f and g are such that they exhibit "regression toward the mean," $\bar{\varepsilon}$ must be close to the mean of ε, with its exact location determined by p and the strength of regression toward the mean.

INTERPRETATION:

- It is not generally possible for players to sanction only when they are "very confident" a norm was violated.
- Players can, however, sanction when they are absolutely sure a norm was violated, or there is no private information

concerning the rate of false positives, i.e., when $\varepsilon = 0$ or ε_i is shared. If ε and ε_i were discrete, players would also have an easier time conditioning on their signal.

Higher-Order Uncertainty

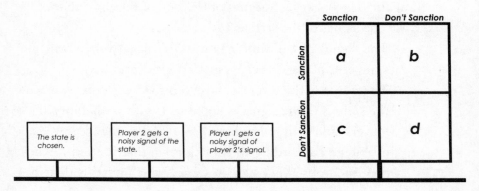

FIRST CASE: NO HIGHER-ORDER SIGNAL

SETUP:

- First the state is chosen. It is 1 with some probability μ and 0 otherwise.
- Player 1 knows the state.
- Player 2 gets a signal of the state as in our earlier model. That is:
  - There are no false positives: when the state is 0, player 2 receives signal 0 with certainty.
  - There are false negatives: when the state is 1, the signal is equal to 1 with probability $1 - \varepsilon$, where ε is the rate of false negatives.

STRATEGY PROFILE OF INTEREST:

- Player 1 plays A if and only if the state is 1. That is, she sanctions whenever she learns of the transgression.
- Player 2 plays A if and only if she gets a signal of 1.

EQUILIBRIUM CONDITIONS:

- $1 - \varepsilon \geq p$.
- $1 - (1 - \mu) / [(1 - \mu) + \mu\varepsilon] \leq p$. This condition will not bind so long as the likelihood of a transgression, μ, is relatively small.

SECOND CASE: ADDING A HIGHER-ORDER SIGNAL

SETUP:

- Everything is the same as before, except . . .
- Player 1 also gets a signal of player 2's signal, with analogous properties. That is:
  - There are no false positives: Whenever player 2 gets a 0, player 1 does, too.
  - There are false negatives: Whenever player 2 gets the signal of 1, player 1 will observe a signal of 1 with probability $1 - \delta$, and 0 otherwise.

STRATEGY PROFILE OF INTEREST:

- Player 1 plays A if and only if her signal of player 2's signal is 1. That is, she sanctions whenever she learns the other has learned of the transgression.
- Player 2 plays A if and only if she gets a signal of 1

EQUILIBRIUM CONDITIONS:

- $1 - \delta \geq p$
- $(1 - \varepsilon) / (\varepsilon + (1 - \varepsilon) \delta) \leq p$

INTERPRETATION:

- Getting a signal of the other player's signal only helps if that signal, too, is observable.

- This signal is most useful when their signal is somewhat noisy. If it's very accurate, then when you learn of a transgression, you can be confident they'll learn of it and can thus directly condition on whether you observed the transgression.

CHAPTER 14

SUBGAME PERFECTION AND JUSTICE

IN CHAPTER 6, WE BRIEFLY RAN INTO THE HATFIELDS AND MCCOYS ON the cusp of their infamous feud. The year was 1878, and the two clans had just finished squabbling over a sow and her piglets. Within a decade, the squabble had escalated into what can only be described as an all-out war, replete with a "massacre" and a "battle." By the time the feud had ended, both families were decimated.[1]

The first few incidents in the feud are now famous but are not really the sorts of things one would think would lead to massacres and battles. In 1880, a full two years after the dispute over the sow, two McCoy brothers got into a gun fight with Bill Stanton, who had testified against the McCoys in the trial over the pig. Stanton was killed, and the McCoy brothers were briefly jailed but soon released when it was decided they had acted in self-defense.

A couple of months later, with tensions between the families already high, a romantic tryst triggered yet more trouble. Johnse Anse Hatfield met Roseanna McCoy at an election event and successfully wooed her (as he had apparently done to just about every other girl in town). Roseanna was daughter to Randall McCoy, head of the McCoy clan. He and her brothers saw Johnse's behavior as an affront to the McCoy family honor, and everyone in the area, including the Hatfields, seemed to side with them on the matter. The McCoy brothers periodically harassed Johnse and at one point even kidnapped him, perhaps with the intent of killing him. Johnse's father, Devil Anse Hatfield, head of the Hatfield clan, sent out a posse and intercepted the McCoys. Devil Anse permanently separated the young lovers, but likely because he viewed the McCoy brothers' behavior as justified—and because Johnse wasn't actually killed—he let them go unharmed.

So how did we go from such isolated incidents to massacres and battles? There were several key points of escalation at which the parties could have backed down and prevented more bloodshed but, well, they very much didn't.

The first of these occurred in 1882. At yet another election event, Devil Anse's brother, Ellison Hatfield, found himself in a drunken brawl with a McCoy brother. Two more McCoy brothers were nearby. They joined the fight, knives were drawn, and Ellison was ultimately killed. The McCoy brothers were immediately arrested, but instead of leaving it at that, Devil Anse took matters into his own hands. He rounded up a party of about twenty and intercepted the constables as they made their way to the jailhouse, forcing the constables to hand over the McCoy brothers. His posse held the McCoy brothers in an abandoned schoolhouse for a few days, as Devil Anse waited to see whether his brother would survive. When Ellison Hatfield died from his wounds, Devil Anse tied the McCoy brothers to pawpaw trees and told his posse to empty their rifles into them.

Devil Anse's vigilante justice was largely perceived as justified, especially since the McCoy brothers had already gotten away with Bill Stanton's murder. Though Devil Anse and his posse were indicted, there was little enthusiasm for arresting them. It helped that the Kentucky lawmen entrusted with arresting the posse were wary of crossing into West Virginia to do the deed. Devil Anse had fostered political connections with West Virginia's governor and helped to get West Virginia's Senator John B. Floyd elected, at one point supposedly appearing at a local election event at the head of an armed party to, uh, persuade voters to vote for Floyd.

Randall McCoy was understandably displeased with this outcome, but it seems he at first tried to live with it. For a while, he
did nothing, and an uneasy truce emerged. Then, in 1887, things
took a very dark turn. Accounts differ, but most say things went
south after the Hatfields planned a raid on the McCoys that was
foiled. Two McCoy women were suspected of alerting the others,
and one of Devil Anse's sons, Cap Hatfield, broke into their house
at the head of a posse and beat the McCoy women badly, whipping
them with cow tails.

This apparently led Randall to change his tune and raise the
stakes dramatically. He joined forces with a distant relative and
friend, Perry Cline, who had powerful political connections in
Kentucky and is sometimes said to have been after Devil Anse's
timber possessions. Cline persuaded Kentucky's governor to rein-
state the Hatfield indictments and, this time, to offer large bounties
to anyone who captured the Hatfields and brought them to Kentucky to stand trial. You've probably gathered that the Hatfields
had no intention of turning themselves in. Upon receiving word,
Devil Anse wrote back: "We . . . do notify you that if you come into
this country to take or bother any of the Hatfields, we will follow
you to hell or take your hide."

But the Hatfields didn't leave it at that. It was their turn to raise
the stakes. On the night of January 1, 1888, they surrounded the
McCoy cabin, set it ablaze, and opened fire (in case it isn't obvious,
this is the "massacre"). Randall McCoy made it out alive, but he
lost a son and daughter in the shoot-out, and his wife was beaten
so badly that she was handicapped for the rest of her life. McCoy's
other children escaped, though several suffered frostbite as they
weren't dressed for the cold.

At this point, Randall McCoy finally made his exit. He moved
the surviving members of his family out of the area to the near-
est town and never returned. He died twenty-six years later, aged
eighty-eight, after being burned by a cooking fire (he really had the
worst luck).

The Hatfields, meanwhile, engaged in an all-out war with bounty hunters and law enforcement. It lasted for much of 1888, even embroiling the governors of both Kentucky and West Virginia. Several Hatfields were finally apprehended in August after losing a shoot-out now known as the Battle of Grapevine Hill. Interestingly, their case ended up in front of the Supreme Court because they weren't properly extradited, but with several key Hatfields in jail and the key McCoy out of the picture, most histories end here, even if the feuding and bounty hunting went on for another decade. Devil Anse himself was never convicted. He lived to age eighty-one.

So, why'd they do it? Why'd Devil Anse have to go and have the McCoy boys executed? He could've just let law enforcement do its thing. Indeed, his brother Ellison pleaded with him to do just that as he lay dying from his wounds. Why did the Hatfields beat the McCoy women? Of course, they could have expected some kind of response from the McCoys. Still, why was Randall so harsh in response? It's not like he couldn't have guessed the Hatfields wouldn't just let him get away with sending bounty hunters after them. But then, why didn't the Hatfields at least tone things down? They clearly knew, when they set off to massacre the McCoys in their cabin on that cold January night, that they would not get away with such an atrocity. Why, in short, did the two sides keep escalating, instead of letting bygones be bygones. So many lives could've been saved!

ON APRIL 13, 2021, ISRAELI POLICE MUSCLED THEIR WAY ON TO THE site of the Al-Aqsa Mosque in Jerusalem and cut the cords to the minaret speakers to prevent the call for prayer from interrupting a speech being given nearby by Israeli president Reuven Rivlin. Coming a day after the start of Ramadan, and on the backdrop of a controversial trial to evict six Palestinian families from the East

Jerusalem neighborhood of Sheikh Jarrah, the incident was seen as an affront and sparked a series of small protests in both Al-Aqsa and Sheikh Jarrah.

Often, such protests eventually die down, and things return to normal. This time, they didn't. Instead, just as happened with the Hatfields and McCoys, a series of escalations led to the worst altercation in almost a decade. It's hard to say exactly who started the escalations and when, but one key incident occurred on May 7 when Israel upped the ante by raiding the Al-Aqsa Mosque compound to confiscate a stash of stones stockpiled by protestors. If cutting the speaker cables three weeks earlier was an affront, the raid, replete with stun grenades and rubber bullets, was an outrage. A second key escalation originated in far-off Gaza. On May 10, Hamas weighed in on the side of the outgunned protestors and issued an ultimatum: Israel was to remove police and military from Al-Aqsa and Sheikh Jarrah by 6 p.m., or else.[2]

Israel ignored them.

Minutes after the deadline passed, Hamas followed through on its promise, sending a barrage of missiles into Israel. Israel responded with artillery fire and airstrikes. Then, over the upcoming days, it leveled several apartment blocks in Gaza, including one housing the Associated Press and another housing the international news network Al Jazeera. Hamas retaliated by intensifying its attacks, sending hundreds of rockets deep into Israel. For eleven days, until both sides finally agreed to a ceasefire that commenced at 2 a.m. on May 21, there were near-constant airstrikes against Gaza and rocket attacks against Israel.[3]

As we did with the Hatfields and McCoys, we could ask why both sides escalated the conflict, instead of letting bygones be bygones. If only Israel hadn't cut the cables to the speakers, if the protests hadn't intensified, if Israel hadn't raided Al-Aqsa, if Hamas hadn't issued its ultimatum and launched its rockets, *etc., etc.,* over two hundred deaths would have been averted. Why, at each stage, did the parties not let bygones be bygones?

Here's another way to pose the question. In these conflicts, both sides had spilt milk. The hog lost by the McCoys was a valuable asset that could've fed Randall's family for an entire winter. The cut speaker cables at Al-Aqsa prevented a call to prayer. But once the milk is spilled, it's what economists would call a sunk cost, so-called because the cost cannot be recovered. Regardless of whether the McCoys or the Palestinians avenge the wrong, they won't get back their hog or the call to prayer. Why cry over that spilt milk?

Indeed, any economist will unsentimentally intone that it's best not to cry over spilt milk—not to get hung up on sunk costs. It doesn't matter how much you've sunk into a project or venture. All that matters is how much it's going to return in the future. And picking a fight—be it with knives or rockets—is hardly a positive net present value proposition. The fight itself is costly, and then, there's the risk it'll lead to an escalation. So why don't the McCoys and the Palestinians and their counterparts ignore sunk costs— ignore past harms—and simply move on?

Indeed, this is precisely the advice you'll often find in self-help texts from popular figures like Oprah and Dr. Phil and even from the health experts at the Mayo Clinic. Here's what the good folks at Mayo have to say on the topic:[4]

> Who hasn't been hurt by the actions or words of another? Perhaps a parent constantly criticized you growing up, a colleague sabotaged a project or your partner had an affair. Or maybe you've had a traumatic experience, such as being physically or emotionally abused by someone close to you.
>
> These wounds can leave you with lasting feelings of anger and bitterness—even vengeance.
>
> But if you don't practice forgiveness, you might be the one who pays most dearly. By embracing forgiveness, you can also embrace peace, hope, gratitude and joy. Consider how forgiveness can lead you down the path of physical, emotional and spiritual well-being.

They even go on to say:

Letting go of grudges and bitterness can make way for improved health and peace of mind. Forgiveness can lead to:

- Healthier relationships
- Improved mental health
- Less anxiety, stress, and hostility
- Lower blood pressure
- Fewer symptoms of depression
- A stronger immune system
- Improved heart health
- Improved self-esteem

Sounds like a pretty good deal. Why, then, do we find forgiveness so counterintuitive? Why don't we treat past harms as sunk? Why don't we let bygones be bygones and just bury the hatchet?

◆

TO ANSWER THESE QUESTIONS, WE'LL INTRODUCE THE REPEATED PUN-ishment game, which is based on the repeated prisoner's dilemma but has some useful tweaks.[5] This new game has two players. In each round of the game, player 1 moves first, deciding whether to transgress or not (we'll sometimes refer to not transgressing as cooperating). Transgressing benefits player 1 but harms player 2. Then, player 2 moves, deciding whether to punish player 1 or not. As usual, after each round, there is some chance the game continues to another, identical round and some chance it ends.[6]

When Devil Anse decided to mount up a posse and kidnap the McCoy brothers, he and his gang placed themselves at risk. The constables could have opened fire on them, or the McCoys could have mounted up a posse and ambushed them. Then there was the fact that the execution might lead to their arrest. Indeed, there are multiple episodes in the feud in which those who set out to punish end up arrested, injured, or killed. We'd like the model to capture this, so we assume that punishing is costly, not just to the player being punished but to the one doing the punishing.[7]

We can learn a lot in this game by considering some simple strategies. The first one is:

- Player 1 never transgresses.
- Player 2 never punishes.

Is this strategy a Nash equilibrium? No. Player 1 benefits by deviating to transgressing because by doing so she gets the gains from transgressing and faces no repercussions. To be an equilibrium, a strategy profile must provide an incentive to cooperate. (If you were paying attention in Chapter 8, you already knew this. There, too, we saw that without the threat of punishment, we cannot deter transgressions in equilibrium.)

So, let's add an incentive to cooperate. Here's a strategy that does that:

- Player 1 never transgresses.
- Player 2 punishes only if player 1 transgressed in the current round.

Now, player 2's threat of punishment can deter transgressions: so long as the harm from being punished is bigger than the gain from transgressing, this strategy profile is Nash. Great!

Not so fast, though. Although player 1 wouldn't transgress, let's think about the hypothetical situation in which she did. We have to do this to make sure this equilibrium is *subgame perfect*, the refinement of Nash that we briefly encountered in Chapter 8. Recall that, to be subgame perfect equilibrium, players must have an incentive to follow through on threats like "I will punish you if you transgress."

So, would player 2 actually punish this (hypothetical) transgression? Hmmm. Punishing is costly. And if he doesn't punish, there are no repercussions: player 1 will simply go back to cooperating in the next round. So, the answer is no, player 2 wouldn't punish. That's a problem. If player 1 anticipates that player 2 wouldn't actually punish, she'll go ahead and transgress.

Here's a strategy profile that fixes this problem.

- In each round, player 1 does not transgress, so long as no prior transgression has gone unpunished; otherwise, she transgresses.
- Player 2 punishes if player 1 transgressed in this round and no prior transgression has gone unpunished; otherwise, he doesn't punish.

The key thing this strategy profile does is introduce repercussions for failing to punish, namely, that player 1 will stop cooperating with player 2 if player 2 fails to punish a transgression.

While this strategy provides these repercussions in a particular way, the lesson is general: a strategy profile must call on players to take advantage of any partner that fails to punish. This additional requirement comes from subgame perfection and, specifically, the requirement that players have an incentive to follow through on their threats. Our third strategy has that incentive in droves, since player 2 will never again see an ounce of cooperation if he doesn't follow through on his threat to punish.

◆

SO, WE ALREADY KNEW THAT PLAYERS MUST PROVIDE AN INCENTIVE for their partners to cooperate. Now we've learned that they must also provide an incentive for their partners to punish, by exploiting a partner who doesn't.

The key implication of this additional requirement is that, in a subgame perfect equilibrium, if player 2 were to try to let bygones be bygones, player 1 would take advantage of him by being less cooperative going forward. That is, in a subgame perfect equilibrium, bygones aren't bygone at all. They're not sunk costs because a transgression that happened yesterday very much affects whether punishment today will be expected, and what failure to punish today might indicate about how transgressions will be treated tomorrow. In repeated games, bygones can matter. Actually, it's stronger than that: in repeated games, bygones must be made to matter. That's the only way to sustain cooperation.

That's why Israel and Palestine can't let past transgressions go unanswered. If the Palestinians don't punish Israel for cutting the speaker cables at Al-Aqsa, Israel will do it again or something like it. In equilibrium, it must. If Israel doesn't punish Hamas for launching rockets, Hamas will do it again. In equilibrium, it must. It's true that this cycle of violence is awful and seemingly counterproductive once we're in it, but it's a necessary part of maintaining peace in the first place!

Likewise, for the Hatfields and McCoys, if they failed to punish a perceived slight, the other shoe would soon drop. We'll never know for sure, but it's even possible that's why the Hatfields planned the foiled 1887 raid on the McCoys. Perhaps after Randall let the indictments against them for the death of his boys lapse (letting a transgression go unpunished), they felt compelled to exploit the McCoys.

For a very clear and well-known case where failing to punish a transgression leads to further exploitation, we can look to the Allies' policy of appeasement toward Hitler on the eve of World War II. The horrors of the First World War had led Allied leaders to adopt this policy in an effort to avoid an escalation of violence and another world war. First, they let Germany's annexation of Austria go unanswered. So Hitler concluded he could be even more aggressive and demanded to annex parts of Czechoslovakia. When England and France capitulated to these demands, too, Hitler remarked to his generals: "Our enemies have leaders who are below the average. No personalities. No masters, no men of action. . . . Our enemies are small fry. I saw them in Munich."[8] Then he invaded Poland.

WILL A MERE APOLOGY DO? IN CHAPTER 5, WE DISCUSSED HOW apologies—mere words—can often have a huge impact because they can change expectations over which equilibrium players are in. (That is, so long as they are explicit, as we saw in Chapter 13.)

Let's take a look at how apologies might work in this model. Suppose that, after choosing whether to transgress, player 1 has an opportunity to apologize. For now, assume the apology is costless—mere words. After this, player 2 chooses whether to punish as before. Would player 2 accept player 1's apology? That is, would he only punish if player 1 transgresses and doesn't apologize?

No. This might be tempting to player 2 since it would mean avoiding having to pay the cost of punishing, but there's a problem. Because player 1 benefits from her transgression, the apology does not sufficiently deter a transgression. She'll simply choose to transgress and then say "Oops, I'm *so* sorry" every time. Moreover, subgame perfection teaches us that a player 2 who does accept a costless apology is bound to be exploited in equilibrium.

If, however, the apology is costly—say it comes with financial compensation or reparations of some kind—then, so long as the cost is greater than the gains from transgressing, it will suffice to deter transgressions (provided player 1 expects to have to pay such a cost to avoid punishment any time she transgresses). Such an apology can be accepted by player 2 without fear of future exploitation.

Fans of the TV Show *The Sopranos* might recall how Tony and Carmela Soprano temporarily split (season 5) because of Tony's repeated, flagrant infidelity. Eventually, the two agree to meet and try to reconcile. A few moments in, Carmela casually drops a hint. "It turns out there is a lot for sale over on Crestview. A little over an acre. I was thinking, I could build a spec house . . . and take my dad in as a partner. I mean, my allowance is what it is, but . . ." Tony gets the drift and agrees to the deal.

"How much is the lot?" he interrupts.

"Six hundred thousand . . ."

"Well, I'll call Ginsberg . . . and have him free up a down payment," says Tony, agreeing to her offer.

"And then?"

"I'll move back in," Tony asserts, before reiterating his promise from a previous conversation that Carmela won't have to deal with his infidelity anymore. The gift has made their reconciliation possible.

So, when must apologies be costly? Whenever the transgressor benefits from the transgression, as Tony did from his infidelities. Contrast this with the case of the cattlemen of Shasta County from Chapter 9 who, though they would typically help to mend a fence, did not pay their neighbors for the damage done by their trespassing cattle. This was possible because the transgression did not benefit the cattlemen. They'd have preferred that their cattle stick together and not wander off and get lost on some neighbors' land. Another example: the United States' accidental killing of Pakistani soldiers—the one that culminated in Hillary Clinton's apology (Chapter 6). The US did not benefit by killing its ally's soldiers. In such cases, there is no need for the apology to be accompanied by reparations; just a clear enough apology that ensures this past transgression doesn't set a precedent for future ones.[9]

DUELS

In 1865, six years before Otto von Bismarck unified Germany, he found himself in the Reichstag, Germany's parliament, attempting to push through his latest budget. A liberal critic, Rudolph Virchow, who was also a scientist and one of the fathers of modern pathology, issued a biting criticism of Bismarck's budget for its excessive military funding. Bismarck was so offended that he challenged Virchow to a duel.

As the story goes, when Bismarck's messenger arrived at Virchow's, he found the scientist-cum-politician at his lab bench. Upon reading the messages, Virchow responded that, as was customary, he would choose the weapon. He pointed to two sausages that lay nearby. One had been loaded with *Trichinella* larvae. The other was safe. Bismarck was to choose which sausage he'd prefer to eat; Virchow would eat the other. Bismarck wisely declined—dying of trichinosis is awful.[10]

Although the story is likely not true, amazingly, it's just the sausage bit that's probably apocryphal (their correspondences suggest Virchow declined and publicly apologized), not the fact that Bismarck willingly challenged a rival to a duel. Indeed, in 1852, he had participated in a duel and survived, and as a student in the 1830s, he had reportedly participated in—and initiated—several more.

Dueling largely fell out of favor over the course of the latter half of the nineteenth century, but in the preceding centuries, it was a common means of resolving disputes. In the West, the practice originated among the Germanic tribes who conquered Western Europe after the collapse of Rome and was the legal means of resolving certain disputes in some early medieval European kingdoms. It was also common among the Vikings. In the High Middle Ages and Renaissance, commoners were increasingly prohibited from engaging in duels, but dueling remained in practice among noblemen and the elite. A relatively large number of prominent individuals took part in duels, including US presidents Andrew Jackson (who narrowly escaped death) and Alexander Hamilton (who, famously, did not). In 1842, Abraham Lincoln agreed to fight a duel with James Shield, but the two were talked out of it. They would go on to be friends; two decades later, Shield would serve as a general under Lincoln during the Civil War.[11]

Although duels were not typically fatal, the death rate was nontrivial. One estimate, from 1685–1716 France, puts it at roughly 4 percent.[12] Of course, injuries were also common. Andrew Jackson carried a bullet from his duel in his chest for the rest of his life, and it caused him enormous pain. You might think this would lead people to decline to duel. However, declining typically led to prompt ostracism by the entire community. As one Mississippi congressman explained, if he were to turn down a challenge, "life will be rendered valueless to him, both in his own eyes and in the eyes of the community."[13]

Duels are, without a doubt, a strange institution. Why require the person who was wronged to pay the same cost—and a hefty one!—as the person who transgressed? And why would anyone willingly pay this cost? Why would such a bizarre institution arise and stick around for so long?

To answer the first question, let's ask what would happen if we could punish a transgressor without paying a cost ourselves. A problem arises: people sometimes benefit from harming a rival, whether he transgressed or not. Maybe they're competing over some land, a job, or a romantic partner. In such cases, there's a perverse incentive: it might be awfully convenient to pretend that one's rival has transgressed so that he can be summarily dispatched under the pretense of being punished. While the threat of being hauled into court might normally deter such behavior, what happens in societies with weak, faraway governments, like those of the early Middle Ages, the Deep South, and the Wild West? In such cases, people need to be deterred from punishing without cause.

Duels do this. If you have to pay the same cost you are incurring on your rival, you aren't going to use it just to cut your rival down to size.

But why would anyone challenge another to a duel, given these costs, even when they had just cause? Well, for all the reasons we spelled out earlier in this chapter: Even if it's costly to punish—4 percent chance of death costly—it is still worthwhile to punish since, otherwise, the transgressor (or others who learn you failed to take the transgression seriously) will continue to transgress against you in the future.

That, then, is our explanation for why duels arose and stuck around. Duels prevent wanton violence, while permitting legitimate deterrence.

MORAL LUCK

A curious incident in the Hatfield and McCoy feud illustrates another puzzling feature of our sense of justice. A short time after
Devil Anse Hatfield executed the McCoy brothers, his son Cap
started a fight with a certain Dr. Reese at a Christmas dance and
got himself shot. For a while, it was unclear whether Cap would
survive. When Anse, who cared for his boy 24/7 until he recovered,
was later asked about the incident, he admitted Cap "was in the
wrong" but added, "If he had killed my boy, I shoulda killed him
to make it square." When that proved unnecessary, Devil Anse
didn't so much as lodge a complaint against Dr. Reese.

How could Anse feel so differently about the two events, when
the difference between them boiled down to luck—a lucky break
for young Cap and an unlucky one for Ellison? Just think, if Lady
Luck had only bestowed the same favor upon Ellison Hatfield as
she had on Cap, decades of feuding and dozens of deaths would
have been averted!

Bernard Williams is usually credited for highlighting the out-
sized role that luck plays on moral intuitions, calling this phenomenon moral luck. Here is Thomas Nagel on the subject:[14]

> If someone has had too much to drink and his car swerves on to
> the sidewalk, he can count himself morally lucky if there are no pe-
> destrians in its path. If there were, he would be to blame for their
> deaths, and would probably be prosecuted for manslaughter. But if
> he hurts no one, although his recklessness is exactly the same, he is
> guilty of a far less serious legal offence and will certainly reproach
> himself and be reproached by others much less severely.

Inspired by Nagel's thought experiment, Fiery Cushman, Anna
Dreber, Ying Wang, and Jay Costa developed the following simple
lab experiment to illustrate how moral luck pervades our moral
intuitions.[15] Subjects were paired up, and one was given ten dollars,

which would be allocated between them according to the roll of one of three possible dice. Player 1 was asked to choose among the three dice, which differed on how generous they are toward player 2: the first die resulted in a one-third chance that player 2 got all the money, the second was evenhanded, and the third resulted in a two-thirds chance that player 2 got all the money. After player 1 made his choice, the die was rolled, and then player 2 was asked if she would like to reward or punish player 1, by up to nine dollars in either direction.

The only thing in player 1's control was the die chosen. Once it had been chosen, the actual outcome was in Lady Luck's hands. Yet, Cushman and his coauthors found that subjects didn't reward or punish their partners only based on the die they chose, but also based on how lucky the roll was. Indeed, they found that subjects hardly responded to their partners' choice of die at all, and mostly based their rewards and punishments on the result of the die roll. Sheesh!

Our explanation for this phenomenon requires that we incorporate higher-order beliefs, which we learned about in the last chapter. Let's walk through why higher-order beliefs might matter in this setting: Put yourself in player 2's shoes and ask yourself what would happen if you thought a transgression had occurred but knew that player 1 did not think so. Would you punish? No. If you did, you'd not only pay the cost of punishing now, but you'd also be unable to have a cooperative relationship in the future, since player 1 would have seen you as punishing when it wasn't appropriate to do so. But if you don't punish, won't player 1 exploit you? Nope! Because player 1 doesn't think a transgression occurred. Just as importantly, the opposite is also true: even if you think no transgression occurred (as Anse did when his boy was the one who picked a fight), you wouldn't want to be seen as failing to deter if player 1 thinks a transgression did occur. Then there's a risk of being exploited!

Now to explain why we, including Devil Anse, take consequences into account, above and beyond intentions, when deciding whether or not to punish, we only need to go back to one of the main messages from the last chapter: consequences, more so than intentions, are liable to be commonly known. Did Dr. Reese harm Cap out of malice or in self-defense? That might be hard to verify. But whether or not Cap dies isn't.

FULL-BLOWN RESPONSES TO
MINUSCULE TRANSGRESSIONS

The Falkland Islands are a small British territory in the Pacific Ocean, three hundred miles off the shore of Patagonia. The population of 3,398 is mostly of British descent, and they make a living primarily in the tourism, wool, or fishing industries. In the early twentieth century, the islands were of some strategic and economic value to Britain, serving as a base from which the Royal Navy maintained control over the South Pacific and providing valuable wool for British industry. However, after World War II ended and wool prices collapsed, the British found that governing the islands was costing them more than it was netting them.

Meanwhile, Argentina insisted that the islands belonged to them. As you might recall from Chapter 5, their claim went all the way back to 1816, when Spain granted Argentina the right to colonies it had abandoned. The British grew increasingly enthusiastic about handing over the islands, and in the mid-1960s, they entered into talks with Argentina about a handoff, though when word of the talks broke, the islanders, who wished to remain part of Britain, successfully lobbied to end them. In the early 1980s, ever mindful of needless spending, British prime minister Margaret Thatcher reopened the talks, though they again fell apart.

Then, on April 2, 1982, Argentina invaded Port Stanley, the largest town on the islands. Thatcher responded by sending in the navy—officially just a task force, though one whose composition of aircraft carriers, nuclear submarines, destroyers, and frigates made it readily apparent that the task at hand wasn't a survey of migrating seabirds. Over the next two and a half months, the UK devoted $1.19 billion and sacrificed 255 British lives to retake the islands, which, ironically, they didn't really want. They've been stuck with them since.

Why would England fight a costly war to defend its territorial claim to the Falkland Islands when they didn't even value that territory? Why, in other words, pay the cost to punish so severely such a trivial transgression?

In 1968, North Korea exhibited the same puzzling behavior. Outraged that the United States was deigning to prune a tree in the demilitarized zone that obscured some critical sight lines, they attacked and killed two American soldiers with axes.[16] Shakespeare fans will recall that in the opening to *Romeo and Juliet*, it takes just thirteen lines to go from "Do you bite your thumb at us, sir?" until the first sword is drawn and just a few more before the scene devolves completely into an outright brawl between the Capulets and Montagues. If you've been paying attention, you'll have noted that the real-life feudists of the Hatfield and McCoy clans were similarly quick to respond to even the slightest assault on their honor by whipping out daggers and pistols and risking their own lives in a brawl or shoot-out.

The key thing that unites all these cases is that the countries or people in question are willing to pay substantial costs to punish even the most trivial transgressions. Why go to war over an island you don't want or a silly tree that's not even on your land? Why risk death over an insult like "I bite my thumb"? That's nuts!

You might intuitively sense that the answer lies with precedent, namely, that if a small transgression goes unpunished then this sets a precedent for larger, more meaningful transgressions. This is true, but it does raise the question of why small transgressions create this precedent. Why is it that we don't expect people to let the small stuff fly and only severely punish more serious transgression?

Once again, we think the key to the answer is to incorporate into the existing framework the lessons from past chapters. This time, the key insight comes from Chapter 13, where we argued that it was difficult to condition on continuous variables in settings in which higher-order beliefs matter—continuous variables like the magnitude of a transgression. This means parties in a dispute will be forced to treat qualitatively similar transgressions as the same, even if the magnitudes of the transgressions are very different—to punish small transgressions as severely as they'd punish large ones.

Here's another example of how our sense of justice tends to be insensitive to continuous variables. In an episode of *Star Trek: The Next Generation* called "Justice," the *Enterprise* discovers a utopian planet whose citizens, the Edo, are "neat as pins, ultra lawful, and make love at the drop of a hat. Any hat." When Captain Picard sends a contingent of his crew to visit this enticing new world, they soon run into a bit of trouble. One of the crew members is frolicking with some locals, trips, and falls into a bed of new flowers. Harming the flowers is forbidden, and soon, two enforcers known as mediators appear. To everyone's surprise, Picard's crew member is sentenced to death. It turns out the way the Edo enforce laws is rather peculiar. Each day, they send mediators to just one zone. Only the mediators know which zone: it's kept a secret from the rest of the population. Anyone who breaks any law in the zone, no matter how small, is punished by death. Since no one wishes to risk death, no one breaks the law. The Edo are quite proud of this system, which successfully deters crime with minimal need for law enforcement, but Picard's crew (and, presumably, the viewer) is horrified. How can the Edo be so sanguine about putting a man to death for a trivial crime?

In fact, the Edo's system of justice makes a lot of sense. Here's why. To deter a transgression, one must consider both the gains from the transgression and the likelihood of getting caught. A diamond thief who is only made to return the diamonds he stole won't be deterred from stealing. If he's caught, he's back where he started, but if he isn't, he's a rich man. Why not take his chances? To sufficiently deter the thief, we must make the punishment bigger than the gains from the crime—sufficiently bigger that even though there's a chance he might get away with the crime, he won't he willing to risk it. How much bigger? That depends on the chances of getting caught. The slimmer the chances, the greater the punishment. This was one of the arguments Gary Becker gave in his 1968 paper "Crime and Punishment: An Economic Approach."

But Becker went further. Catching thieves requires nontrivial resources: a big police force with smart detectives who have access to fast cars, big guns, surveillance, DNA tests, and whatever else it is that police and detectives use to enforce the law. Instead of doing all this, Becker suggested that we could simply increase the punishment. That would make it possible to reduce the costs of enforcing laws, without reducing the effectiveness of law enforcement. This is precisely the ingenious system that the Edo came to employ on their utopian planet.

Becker's ideas have had a big influence not just on the Edo but also on Earth. From 1976 to 1999, these ideas were taught to nearly half of all US judges at the Manne Economics Institute for Federal Judges. In a paper by Elliott Ash, Daniel Chen, and Suresh Naidu that considered the impact of these training programs, the authors showed that attendees increased the severity of criminal sentences by 10 to 20 percent and even judges who themselves didn't attend the Manne workshops were influenced by them when they sat on the bench with judges who had attended.[17]

However, as ingenious as Becker's solution is, to most of us it feels wrong, particularly in cases like the one highlighted in the *Star Trek* episode, in which the chances of getting caught are low but the crime is trivial. In such cases, the required punishment feels to most of us to be disproportionate to the crime. This intuition might seem puzzling until we recognize that our intuitions regarding justice are incapable of being sensitive to continuous variables like the chances of getting caught.

Quirks of our sense of justice—like moral luck and insensitivity to continuous variables like the magnitude of a transgression and the likelihood of being caught—are sometimes interpreted as evidence that our sense of justice isn't intended to deter transgressions. After all, if you wanted to deter transgressions, you would condition punishments on the thing people actually have control over, like which die they chose, whether they were driving drunk, or whether they shot your son. You would likewise want to take into account how likely someone is to get caught in determining their punishment. And there would be no need to make mountains out of mole hills as England did with its Falkland Islands invasion.

However, our interpretation is that while our sense of justice might ideally ignore luck and take into account the likelihood of being caught and the magnitude of the transgression, it is constrained from doing so because of higher-order beliefs. That doesn't mean the purpose of our sense of justice isn't to deter transgressions, just that we must also account for the quirky influences of higher-order beliefs.

SO THOSE ARE A FEW QUIRKY FEATURES OF OUR SENSE OF JUSTICE that we think are well explained by subgame perfection, illustrated via a simple model of how punishment works in dyadic interactions—the repeated punishment game.

And that's the last game theory chapter we have for you.

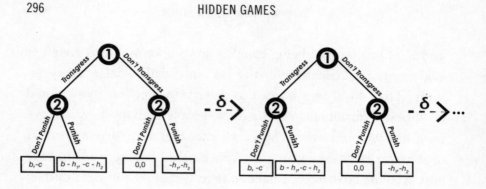

SETUP:

- There are two players. In each round, the punishment game is played. With probability δ they continue on to the next period. Otherwise the game ends.
- In the punishment game, player 1 first chooses whether to transgress or not. Then, player 2 chooses whether to punish or not. Transgressing benefits player 1 by $b > 0$, and costs player 2 $c > 0$. Punishment costs player 2 $h_2 > 0$ and harms player 1 by $h_1 > 0$.
- In this model, players can observe everything that's happened in the past.

SOME STRATEGY PROFILES OF INTEREST:

- Not Nash: in each round, player 1 doesn't transgress and player 2 doesn't punish.
- Nash: in each round, player 1 doesn't transgress and player 2 punishes if and only if player 1 transgressed in the current round.
- Subgame perfect:
  - Player 1 doesn't transgress so long as a defection has never gone unpunished; transgresses otherwise.
  - Player 2 punishes if player 1 transgressed in this round and no transgression has gone unpunished; doesn't punish otherwise.

EQUILIBRIUM CONDITIONS:

- The "Nash" strategy profile is a Nash equilibrium so long as the harm to player 1 from being punished is greater than the gains from transgressing, $h_1 \geq b$. It is never a subgame perfect Nash equilibrium.
- The "subgame perfect" strategy profile is a subgame perfect equilibrium so long as $h_1 \geq b$, and the cost of punishing isn't too large, relative to the harm from future transgressions, i.e., $h_2 \leq \delta c / (1 - \delta)$.

INTERPRETATION:

- As in the repeated prisoner's dilemma from Chapter 8, transgressions are deterred in equilibrium via the threat of punishment.
- However, for a strategy to be a subgame perfect equilibrium, punishment must also be incentivized. For instance, in the "subgame perfect" strategy, if player 2 were to let a transgression go unpunished (to "let bygones be bygones"), player 1 would presume that player 2 won't punish in the future, and "takes advantage of him" by transgressing more often in the future.

CHAPTER 15

THE HIDDEN ROLE OF PRIMARY REWARDS

IF YOU'VE MADE IT THIS FAR, THEN YOU'VE PROBABLY GOTTEN A HANG of what this book is about. Pose a puzzle or two or three. Build a stylized game theory model that uncovers the hidden rewards at play. Interpret the model. Lather, rinse, and repeat enough times, and the game theory will explain a dizzying array of puzzles across an impressive range of domains—rights, aesthetics, ethics, altruism, spin, *etc., etc.*

Game theory has proved to be an indispensable tool as we've gone through this exercise. Take, for instance, costly signaling. Did you know that in the first edition of *The Selfish Gene*, Richard Dawkins actually came out against costly signaling as an explanation for ostentatious long tails? He can be forgiven for doing so. Why on Earth would doing things that are counterproductive— that make it easier for foxes to eat you—make you more desirable? Hogwash! But Nash says otherwise, *if* these wasteful things are more wasteful for the less fit, wealthy, clever, or well-connected types. Nash teaches us that, in fact, this waste provides valuable information! And that those who waste are ultimately compensated for doing so.

Or what about the repeated prisoner's dilemma? Did you know that Charles Darwin himself puzzled over that one and failed to come up with the answer? How can costly cooperation actually be in the interest of the cooperator? Well, Nash says that one way is for us to punish those who don't cooperate. Then, subgame perfection taught us that we must also punish those who don't punish when they should. And from all this we gleaned a host of interesting implications about observability, expectations, higher-order beliefs, categorical norms, and so on.

But it wasn't just Nash that helped us disentangle the incentives at play. We also had to be sure we were focused on the *right* incentives—what we called primary rewards. In the costly signaling chapter, we couldn't focus on the fact that people get pleasure from looking at mosaics, wearing fancy clothes, or drinking *grand vin de Bordeaux*. We had to focus on the primary reward of being accepted as a mate, romantic partner, friend, business associate, and so on. In the altruism chapter, we couldn't focus on what people tell you they care about ("I just want to do good!") or the warm, fuzzy feelings they get from doing good. We could only use game theory once we'd focused on the social returns that accrue from doing good. Yes, game theory came in handy, but it wasn't acting alone. It was working in tandem with this focus on primary rewards.

The message we want to leave you with is that focusing on primary rewards is valuable. Invaluable. To hammer home this point, we'll finally turn to the very first puzzle we posed, about passions. This time, though, we won't be using game theory—just a crystalline view of the primary rewards at play.

◆

WHEN WE OPENED THE BOOK WITH A DISCUSSION OF PASSIONS, WE asked: Why do some people become extremely passionate about certain things, like playing chess or the violin? Why do they become passionate about those particular things?

To address this question, it helps to ask what function passions are designed to serve. We think the answer is: passions are designed to motivate us to invest in developing skills and expertise that are likely to reap material and social benefits later on. Fischer, Ramanujan, and Perlman amassed great respect, fame, and legacy. Perlman and Fischer also amassed a reasonable fortune. Perlman even met his wife through music. Ledecky is still young, yet by the time she graduated college, she was already a household name and had amassed an estimated fortune of $4 million.[1]

However, to explain why not everyone is maximally passionate about everything, we need to account for the cost involved in being passionate. What might that be?

The cost, we suspect, is that passions eat up time. Fischer, Ramanujan, and Perlman spent the bulk of their waking hours focused on chess, math, or violin. Ledecky spent fifteen to twenty hours a week in the pool, plus a bunch more at the gym, while a full-time student. All those hours are costly because they could be used for something else. Fischer could have used them to do his homework or learn to socialize. Ramanujan might have spent them with his wife and children or caring for his failing health. Perlman could have spent them playing in the street with the rest of the neighborhood children, whom he could hear from the window of his practice room as he played the same scales and passages over and over and over again. Ledecky could have used them to do basically anything—to take more classes, go to more parties, join a club or two, and so on.

So that's the trade-off: passions cost time but can lead to substantial rewards in terms of respect, fame, legacy, romantic opportunities, and, sometimes, money.

Notice that this trade-off does not consider the *psychological* costs and benefits of passions—how great it feels to be passionate, for instance. If you've gotten this far, you already know those are the kinds of proximate experiences that we mean to explain. Our explanation will, therefore, involve only the kinds of costs and benefits that guide what we learn to find interesting or pleasing or meaningful, like the time lost and the fame or resources acquired. These are the kinds of things that our learning processes respond to. These are primary rewards.

As you probably recall, to distinguish between psychological costs and benefits—the kind that we might consciously experience—from the kinds of costs and benefits that shape what we learn to like and feel, we introduced the term primary rewards. These are the kinds of costs and benefits we think help us understand how passions work.

Let's talk about passions a little bit more to illustrate some of the benefits of thinking in terms of primary rewards.

So far, our account of passions is simple and somewhat obvious, at least *ex post*, but it does provide quite a few less-than-obvious predictions. Here are four of them.

The first is that you won't develop a passion for things that aren't socially valued. No use investing the proverbial ten thousnd hours if nobody ends up caring. Could Fischer have been as good at go as he was at chess? Perlman at the recorder as he is at the violin? Quite possibly, but go wasn't as respected as chess in 1950s New York, and the recorder isn't as prestigious as the violin. No matter how many hours Fischer and Perlman would have invested in go and the recorder, it would not have yielded the respect, fame, and legacy that made their passions for chess and the violin worthwhile.

Our second prediction is that whether you develop a passion for one thing doesn't just depend on how good you are at that thing but also on how bad you are at other things. Perlman, for instance, highlights that his disability played an important role in the development of his passion for the violin. He jokes that, as a child, his parents encouraged him to practice because they probably thought to themselves, "Hey listen, you have a talent, use it, because you're not going to be a tennis player."[2] Being unable to play tennis of course made Perlman less likely to develop a passion for tennis. It might also have made him more passionate about the violin by reducing the cost of spending all those hours on the violin.

Our third prediction: when the "economics of superstars"[3] are at play, so are passions. There are some activities and professions in which superstars are respected and paid much more than everyone else. In law, a small number of lawyers make it to partner, and they do way better than other lawyers. In sports, a few hundred players make it to an elite league like the NBA or the Premier League and become both wealthy and household names; the rest make relatively little and are forgotten. In art, it's the same story: a tiny number of Perlmans do well for themselves, while the rest starve. When the rewards go to a few, an enormous investment of time is needed to succeed, and it's best to go all in, if you're going to go in at all. Passions are well suited for such settings: if you develop a passion, you'll put in the necessary hours, and if you don't, you won't.

And, finally, our fourth prediction is that you are only liable to become passionate if you have a decent chance of being the best. For a young Michael Jordan or LeBron James, a passion for basketball might pay off one day. For those of us who can't reach the top shelf in the kitchen without climbing onto the countertop? No chance. Even if, as kids, we enjoyed playing basketball, our passion for it was likely supplanted by a love of reading or of doing geometry proofs. Jordan, James, Fischer, Ramanujan, Perlman, Ledecky, and Biles are all remarkably talented people, and while we can't account for where those talents come from, we can say that the folks who are more talented—who stand a chance of becoming superstars—are more likely to develop a passion.

◆

OFTEN WHEN WE REVERSE ENGINEER THE PRIMARY REWARDS BEHIND one phenomenon, we find all sorts of previously hard-to-explain details that fit as well. Such parsimony gives us even more confidence that our explanation is on the right track. For passions, there's a constellation of social psychology phenomena that end up seeming to also fit the explanation we just gave for passions and similarly benefiting from identifying their function.

For instance, in *Man's Search for Meaning*, Auschwitz survivor Viktor Frankl argues that the key thing that distinguished campmates who perished from those who survived was a sense of meaning. Those who had it survived; those who didn't succumbed.

Frankl went on to counsel people to find a sense of meaning in life, founding an entire school of therapy around this principle known as logotherapy. But what gives us a sense of meaning? And why can't we just get meaning from whatever it is that we're already doing?

Our broad-strokes answer suggests that we'll get a sense of meaning from things that are socially rewarded. Frankl got meaning from thinking about publishing his work when he would get out of Auschwitz, being there for his family, and eventually offering psychological counsel to others in search of meaning. All of these were indeed likely to be socially rewarded. Frankl didn't get meaning from the scraps of food he managed to scrounge. The food was a reward in and of itself and not one that carries with it any further social rewards. We similarly don't find meaning in indulging in a bar of chocolate or in completing chores—these come with their own rewards, which aren't social, and we don't require a sense of meaning to be motivated to pursue them.

The next phenomenon we turn our attention to is called grit, by which researchers mean people's willingness to persevere in the pursuit of long-term goals. Angela Duckworth has documented that grit is one of the best predictors of success in a diversity of scenarios, including grade school, Ivy League colleges, West Point, and even the National Spelling Bee.[4] As we did with passions and meaning, we wonder: Why are some people grittier than others? And why isn't it easier to develop?

Our broad-strokes answer suggests that grit, like passions, requires an investment of time and resources that could be devoted to other things. It's particularly useful for skills that take a while to acquire—like those necessary to excel in grade school, the Ivy Leagues, and West Point—and likely also helps in domains where superstars get the bulk of the glory, like a spelling bee or the Olympic pool. Will a particular child or teen acquire grit? That depends. How likely are they to excel? That rests on their talents, as we already discussed, but also on their ability to continue to invest, which might be hampered if, say, they have to hold down a job during high school. If they do excel, will they receive rewards? That depends on what's socially valued, as we already saw, but also on whether they have the relevant opportunities—the social network that might lead to an interview where they can show off the skills they've acquired, the support necessary to prepare for that interview, the ability to travel to it, and so on. Then there's the question of how good they are at other things. If they're really gritty about homework, that can be great but not as great if it removes them from contention for an athletic scholarship or pulls the rug out from under a vibrant and healthy social circle. All these factors can help us understand why grit isn't more forthcoming and perhaps even to be more sensitive to the particulars of a kid's situation before advocating that he or she become gritty.

Next, we turn to a series of classic studies by Edward Deci that demonstrate that people often lose motivation to do work when they are paid for it. This is known as crowding out of intrinsic motivation. In Deci's most famous experiments, subjects were asked to solve a relatively interesting puzzle. Then the experimenter used a made-up excuse to leave subjects with the puzzle for a few minutes and observed whether subjects continued to play with the puzzle even though they were no longer expected to. There were two treatments: in the control, subjects were paid just to show up, whereas in the treatment group, they also received a dollar for solving the puzzle correctly. Deci found that subjects who were paid to solve the puzzle were less likely to continue to play with it after the experiment had seemingly ended. Deci replicated his results in a field experiment involving the student newspaper, finding that employees of the paper who were paid to write headlines were less motivated than those who weren't paid, even two months after the payments had ended.[5] Other researchers have since replicated these results with preschoolers and high schoolers, as well as adults trying to lose weight, stop smoking, or remember to wear a seat belt. Why would financial incentives have this counterintuitive effect on people's intrinsic motivation?

Roland Benabou and Jean Tirole have suggested an explanation for this phenomenon that ties in with our broad-brush explanation for passions.[6] If someone has to pay you to do something, that's a good signal that it's only worth doing if you get paid to do it and that it doesn't lead to social rewards. If this thing were worth doing for the social rewards, there'd be no point in paying you to do it! So, it's better not to become (or stay) intrinsically motivated to pursue this thing, lest you continue to pursue it even after the financial payments cease.

A similar explanation fits a finding about intrinsic motivation and feedback. In the early 2000s, a team of clever researchers set out to demonstrate an admittedly well-accepted fact: that people lose motivation when they feel their work isn't properly recognized. In their experiments, they asked subjects to perform somewhat tedious tasks, like counting the number of times the letter *s* appeared twice in a row in a random sequence of letters or using Lego blocks to build action figures known as Bionicles. They paid all subjects the same amount for the number of pages they completed or the number of figures they built, but they varied the way in which they recognized the work. For some subjects, they recognized their work by reviewing and carefully filing their sheets or by prominently displaying all the completed Bionicles side by side. For some subjects, they ignored subjects' work product or, worse, shredded the pages or disassembled the Bionicles as soon as subjects submitted them. Consistently, they found that even though subjects were paid the same, those who received recognition stuck it out longer—about 50 percent longer—than those whose work product was ignored, while those whose work was destroyed stuck it out far less.

Why is recognition so important in determining our motivations? Precisely because such feedback is a helpful signal as to whether there will be sufficient social rewards for intrinsic motivation to pay off.

The next psychological phenomenon we turn to is flow. Psychologists use the term to describe the mental state people are in when they are totally absorbed in a task, losing track of time and becoming oblivious to their surroundings. The term was invented in the mid-1970s by the psychologist Mihály Csikszentmihályi, who studied flow with the goal of learning how to induce it in the workplace so that workers would enjoy their jobs more and be more productive. Csikszentmihályi found there were three key conditions needed to induce flow: clear goals, immediate feedback, and a balance between opportunity and capacity—that is, the task needs to be difficult enough to be challenging but not so difficult that it is impossible, given the skills of the person attempting it. Why do we experience flow when and only when these particular conditions are met?

Being in the zone means not really considering outside options for our time, which we do when we are bored. It's also quite pleasant, which presumably motivates us to get into this state and to stay in it once there. Being challenged and succeeding are precisely the optimal conditions for acquiring human capital, so it makes a ton of sense that we would only feel flow when those conditions are met.

Finally, these arguments about the role of feedback in shaping grit, flow, and intrinsic motivation connect with a classic line of research on learned helplessness. In the late 1960s, the psychologist Martin Seligman subjected dogs to cruel experiments involving a series of random electric shocks.[7] He placed a lever before the distressed dogs, who tried pressing the lever to end the shocks, but it did nothing. Eventually, the dogs lay down and simply whimpered helplessly whenever a shock came.

After some time, Seligman started a new experiment. Again, he shocked the poor dogs, but this time, he made it so that if the dogs pressed the lever, it ended the shock. He brought the dogs who had participated in his previous experiment back in for this new experiment and also brought in some new dogs, who had never participated before. If you were paying attention in Chapter 2, then you've probably guessed that the dogs quickly learned to press the lever and end the shock, and this is what happened—but only for the new dogs. The ones who participated in Seligman's previous experiment again lay down and whimpered helplessly, without even trying to end the shocks with the use of the lever.

Such learned helplessness might initially seem counterproductive, but in fact, it's quite reasonable. Seligman's dogs had received ample feedback that their efforts were in vain and, not unlike the subjects whose Bionicles had been dismantled before their eyes, lost motivation.

Seligman's experiments with dogs have since been replicated in other animals like rats. They've also been used to explain why when people from disadvantaged backgrounds fall upon an unusual opportunity they aren't as motivated to capitalize on it as one might expect. Again, this might seem counterproductive in a particular case, but overall, it's consistent with the idea that people are likely to have intrinsic motivation, grit, passion, and so on when there is a likelihood of success, but when chances of success are low, these qualities are likely to be absent.

◆

WHAT WE LIKE ABOUT THIS ANALYSIS OF PASSIONS IS JUST HOW MUCH you can explain with so little. All we really needed to do was to think through the costs and benefits, calculated in terms of primary rewards. These were straightforward. Various rewards accrue from building up skills. This must be weighed against the time, attention, and monetary cost of building up that skill. Along the way, we encountered some basic economic principles like comparative advantage, economies of superstars, and the theory of human capital investment, but these were not hard or even crucial.

What *was* crucial was keeping straight what counts as a cost and a benefit—making sure our accounting system was the right one, that we were counting primary rewards and not other things, like how good or meaningful or boring it feels to engage in the task at hand.

Our analysis of passions is just one case study meant to illustrate a more general point. This book has focused on game theory, which is a powerful tool for analyzing primary rewards. That tool has solved its share of mysteries and has more to solve—there are plenty more games that we didn't get to and that haven't been invented yet. But it's the *hidden* part—the hidden accounting system of primary rewards—that's really powerful. Once we focus our attention on it, there are lots more mysteries we can solve.

ACKNOWLEDGMENTS

WE THANK BETHANY BURUM FOR THOROUGHLY REVIEWING EACH chapter, Eli Kramer for carefully fact-checking our claims, and Andrew Ferdowsian for reviewing our figures and legends. All remaining errors are ours, not theirs. We thank Adam Bear, Dan Becker, Rob Boyd, Brad Leveck, Andy McAfee, Patrick McAlvanah, Helena Miton, Cristina Moya, Michael Muthukrishna, Lionel Page, Kyle Thomas, Jonathan Schulz, Henry Towsner, and Dan Williams for providing thoughtful feedback on early drafts. We also thank Christian Hilbe, Martin Nowak, Sandy Pentland, and Dave Rand, for their mentorship and financial support as we worked on this project. Finally, we thank the many students, colleagues, family members, and friends who, over the years, have helped us develop these ideas and hone our arguments.

NOTES

CHAPTER 1: INTRODUCTION

1. Rhein, John von. "'Itzhak' an Intimate Film Portrait of Violinist Perlman." *Chicago Tribune*, 3 Apr. 2018, www.chicagotribune.com/entertainment/ct-ent-classical-itzhak-0404-story.html.

2. "Pablo Picasso." *Encyclopedia Britannica*, www.britannica.com/biography/Pablo-Picasso. Accessed 9 Aug. 2021; and "15 Pablo Picasso Fun Facts." Pablo Picasso: Paintings, Quotes, & Biography, www.pablopicasso.org/picasso-facts.jsp. Accessed 9 Aug. 2021.

3. "GiveWell's Cost-Effectiveness Analyses." GiveWell, www.givewell.org/how-we-work/our-criteria/cost-effectiveness/cost-effectiveness-models. Accessed 9 Aug. 2021.

4. Boyle, Kevin J., et al. "An Investigation of Part-Whole Biases in Contingent-Valuation Studies." *Journal of Environmental Economics and Management*, vol. 27, no. 1, 1994, pp. 64–83, doi:10.1006/jeem.1994.1026.

5. If you're not already familiar with this literature, we recommend Kahneman and Thaler's books, as well as those by Dan Ariely.

6. If you're not familiar with this argument, a good place to start might be Richard Dawkins's *The Selfish Gene* (Oxford, UK: Oxford University Press, 1976) or Steven Pinker's *How the Mind Works* (New York: W. W. Norton, 1997). For more—and more interesting—examples of how evolution shaped our preferences, and some implications this has for policy, check out Robert H. Frank's books.

CHAPTER 2: LEARNING

1. Here's a link to the video: www.youtube.com/watch?v=TtfQlk GwE2U.

2. Reinforcement learning is far more sophisticated than we let on in this chapter. In particular, it makes use of a lot of knowledge that has been preprogrammed into us by evolution. We use data efficiently by structuring it, collecting and using only what's valuable and only when it is relevant. We have reasonable first guesses (priors). We do all this even as infants and toddlers. All of this makes us very efficient, not just at reinforcement learning but at learning in general. Here are some papers and books that make this point: Tenenbaum, J. B., C. Kemp, T. L. Griffiths, and N. D. Goodman. "How to Grow a Mind: Statistics, Structure, and Abstraction." *Science*, vol. 331, no. 6022, 2011, pp. 1279–1285; Gopnik, Alison, Andrew N. Meltzoff, and Patricia K. Kuhl. *The Scientist in the Crib: Minds, Brains, and How Children Learn*. New York: William Morrow, 1999; Schulz, Laura. "The Origins of Inquiry: Inductive Inference and Exploration in Early Childhood." *Trends in Cognitive Sciences*, vol. 16, no. 7, 2012, pp. 382–389; Gallistel, C. R. *The Organization of Learning*. Cambridge, MA: MIT Press, 1990; and Barrett, H. Clark. *The Shape of Thought: How Mental Adaptations Evolve*. Oxford, UK: Oxford University Press, 2014. This work is super cool, but importantly, it doesn't undermine the main lesson of this chapter: that learning is pretty good at leading us to optimize. On the contrary, it reinforces this point, since it makes learning an even more efficient optimizer.

3. Many of the examples of social learning presented in this chapter are adapted from Joe Henrich's *The Secret to Our Success* (Princeton, NJ: Princeton University Press, 2015), which we enthusiastically recommend.

4. Zmyj, Norbert, David Buttelmann, Malinda Carpenter, and Moritz M. Daum. "The Reliability of a Model Influences 14-Month-Olds' Imitation." *Journal of Experimental Child Psychology*, vol. 106, no. 4, 2010, pp. 208–220.

5. Jaswal, Vikram K., and Leslie A. Neely. "Adults Don't Always Know Best: Preschoolers Use Past Reliability over Age When Learning New Words." *Psychological Science*, vol. 17, no. 9, 2006, pp. 757–758.

6. VanderBorght, Mieke, and Vikram K. Jaswal. "Who Knows Best? Preschoolers Sometimes Prefer Child Informants over Adult

Informants." *Infant and Child Development: An International Journal of Research and Practice*, vol. 18, no. 1, 2009, pp. 61–71.

7. See Gergely, György, Harold Bekkering, and Ildikó Király. "Rational Imitation in Preverbal Infants." *Nature*, vol. 415, no. 6873, 2002, p. 755; and Gellén, Kata, and David Buttelmann. "Fourteen-Month-Olds Adapt Their Imitative Behavior in Light of a Model's Constraints." *Child Development Research*, vol. 2017, 2017, pp. 1–11, doi:10.1155/2017/8080649. For a helpful summary, see: Harris, Paul L. *Trusting What You're Told*. Cambridge, MA: Harvard University Press, 2012.

8. Of course, learning can also lead to systematic errors, which may be interesting in their own right. We again recommend Joe Henrich's *The Secret to Our Success: How Culture Is Driving Human Evolution, Domesticating Our Species, and Making Us Smarter* (Princeton, NJ: Princeton Univeristy Press, 2015) for great examples of this. In our book, though, we will be focusing on what happens when learning does what it is designed to do—help us adapt.

9. This is not to say that conscious insights based on a causal understanding play no role in innovations like these and in optimization more generally. From our standpoint, this doesn't undermine the use of game theory. On the contrary, insight likely helps to accelerate and improve optimization and makes us more confident in using game theory. Its use to analyze conscious optimization is just not as counterintuitive to people.

10. Billing, Jennifer, and Paul W. Sherman. "Antimicrobial Functions of Spices: Why Some Like It Hot." *Quarterly Review of Biology*, vol. 73, no. 1, 1998, pp. 3–49.

11. A growing army of economists led by Nathan Nunn have employed fantastic empirical methods to document the ways cultures lag, instilling preferences that may have been functional at one point but remain embedded long thereafter. If this literature is of interest to you, then, in addition to Nunn's papers, which we enthusiastically recommend, here are some others to have a look at: Madestam, A., D. Shoag, S. Veuger, and D. Yanagizawa-Drott. "Do Political Protests Matter? Evidence from the Tea Party Movement." *Quarterly Journal of Economics*, vol. 128, no. 4, 2013, pp. 1633–1685, doi:10.1093/qje/qjt021; Giuliano, P., and A. Spilimbergo. "Growing Up in a Recession." *Review of Economic Studies*, vol. 81, no. 2, 2013, pp. 787–817, doi: 10.1093/restud/rdt040; Malmendier, Ulrike, and Stefan Nagel. "Depression Babies: Do Macroeconomic Experiences Affect Risk-Taking?"

SSRN Electronic Journal, 2007, doi:10.2139/ssrn.972751; and Alesina, Alberto, and Nicola Fuchs-Schündeln. "Good-Bye Lenin (or Not?): The Effect of Communism on People's Preferences." *American Economic Review*, vol. 97, no. 4, 2007, pp. 1507–28, doi:10.1257/aer.97.4.1507.

CHAPTER 3: THREE USEFUL DISTINCTIONS

1. Rozin, P., and D. Schiller. "The Nature and Acquisition of a Preference for Chili Pepper by Humans." *Motivation and Emotion* vol. 4, no. 1, 1980, pp. 77–101.

2. Uri Gneezy has a substantial body of work showing the perverse effects financial incentives can sometimes have. For more examples, check out his book with John List, *The Why Axis: Hidden Motives and the Undiscovered Economics of Everyday Life*. New York: PublicAffairs, 2013.

3. Here's a link to the video: www.youtube.com/watch?v=MO0r930Sn_8.

4. We found this earnest discussion on StackExchange. Here's the link: https://scifi.stackexchange.com/questions/172890/in-universe-explanation-for-why-is-the-tos-era-enterprise-is-more-austere-than-t.

5. Davidson, Baruch S. "Why Do We Wear a Kippah?" Chabad.org. www.chabad.org/library/article_cdo/aid/483387/jewish/Why-Do-We-Wear-a-Kippah.htm. Accessed Aug. 9, 2020.

6. We found these quotes on Quora here: www.quora.com/What-do-Protestants-think-of-the-Pope.

CHAPTER 4: SEX RATIOS

1. A complete copy of *The Descent of Man* is available at the following link: http://darwin-online.org.uk/content/frameset?itemID=F937.1&viewtype=text&pageseq=1.

2. Fabiani, Anna, Filippo Galimberti, Simona Sanvito, and A. Rus Hoelzel. "Extreme Polygyny Among Southern Elephant Seals on Sea Lion Island, Falkland Islands." *Behavioral Ecology*, vol. 15, no. 6, 2004, pp. 961–969, doi:10.1093/beheco/arh112.

3. Berger, Michele. "Till Death Do Them Part: Eight Birds That Mate for Life." *Audubon*, 2 Oct. 2017, www.audubon.org/news/till-death-do-them-part-8-birds-mate-life.

4. Edwards, A. W. F. "Natural Selection and the Sex Ratio: Fisher's Sources." *American Naturalist,* vol. 151, no. 6, 1998, pp. 564–569, doi:10.1086/286141. For you science nerds, here's an odd bit of historical trivia. Darwin actually included this answer in the first edition of *The Descent of Man,* but he got cold feet and removed the answer, along with the whole discussion of sex ratios, from the second edition. For a long time, people believed Fisher came up with the answer himself, but we now know that Fischer owned a copy of the first edition of *The Descent of Man* and suspect that he thought Darwin's logic was so commonly accepted among other biologists that he did not bother to provide a citation when repeating Darwin's argument. To be consistent with convention, we'll go ahead and call it Fisher's answer despite this.

5. Trivers, Robert L., and Hope Hare. "Haploidploidy and the Evolution of the Social Insect." *Science,* vol. 191, no. 4224, 1976, pp. 249–263.

6. Herre, Edward Allen. "Sex Ratio Adjustment in Fig Wasps." *Science,* vol. 228, no. 4701, 1985, pp. 896–898.

7. Trivers, R. L., and D. E. Willard. "Natural Selection of Parental Ability to Vary the Sex Ratio of Offspring." *Science,* vol. 179, 1973, pp. 90–92.

8. In mammals, being successful tends to have a small effect on sex ratios. In birds and some insects, the effect is stronger, perhaps because a female can more easily adjust the sex of her offspring based on her health, status, or environmental conditions. For birds, this is because the female carries both X and Y chromosomes (as opposed to mammals, in which the female carries only Xs). For the insects, it's because unfertilized eggs grow into females while fertilized ones grow into males, and the mother stores sperm from her mate and only fertilizes some of the eggs. What about humans? Amusingly, CEOs and presidents are much more likely to have boys. And our sex ratio tracks the S&P 500: when the market booms, there's a glut of boys. But, admittedly, there aren't that many CEOs, presidents, and stock market booms and busts, so this could just be statistical noise. Nonetheless, there are a host of less exciting studies that show small but robust effects. We thank Carl Veller for his help summarizing this literature and recommend his paper (with David Haig and Martin A. Nowak) "The Trivers–Willard Hypothesis: Sex Ratio or Investment?" *Proceedings of the Royal Society B: Biological Sciences,* vol. 283, 11 May 2016, https://royalsocietypublishing.org/doi/10.1098/rspb.2016.0126.

CHAPTER 5: HAWK-DOVE AND RIGHTS

1. We learned about this argument from Nobel Prize–winning game theorist Roger Myerson. See his paper "Justice, Institutions, and Multiple Equilibria," available at http://home.uchicago.edu/~rmyerson /research/justice.pdf.

2. Smith, J. Maynard, and G. R. Price. "The Logic of Animal Conflict." *Nature*, vol. 246, no. 5427, 1973, pp. 15–18, doi:10.1038/2460 15a0.

3. Davies, Nicholas B. "Territorial defence in the speckled wood butterfly (Pararge aegeria): the resident always wins." *Animal Behaviour*, vol. 26, 1978, pp. 138–147.

4. Shaw, Alex, Vivian Li, and Kristina R. Olson. "Children Apply Principles of Physical Ownership to Ideas." *Cognitive Science*, vol. 36, no. 8, 2012, pp. 1383–1403, doi:10.1111/j.1551-6709.2012.01265.x.

5. To answer such questions, one would likely need to consider a number of factors beyond those characterized by the hawk-dove game. For instance, efficiency likely plays a large role: various pressures will likely lead inefficient systems of rights to be replaced by more efficient ones. Rights may also be constrained to be consistent with the values held and proclaimed within a culture. There may be some randomness that sneaks in and then persists in cases where other pressures don't kick in.

6. Rogin, Josh. "Inside the U.S. 'Apology' to Pakistan." *Foreign Policy*, 3 July 2012, foreignpolicy.com/2012/07/03/inside-the-u-s -apology-to-pakistan.

7. You might be aware of the controversy surrounding this TED talk. Some of the results Dr. Cuddy cited in the talk (those on the effects of power poses on hormones) did not replicate. However, we think the effect of power poses on subjective feelings of empowerment did replicate. Moreover, the techniques she advocates are commonly used by athletes and actors before big games and shows. Most importantly for our sake, the fact that the talk has so many views—it is the third-most popular TED talk ever and has roughly the same number of views as a top single on YouTube—speaks to the demand for such advice.

8. Moskowitz, Clara. "Bonding with a Captor: Why Jaycee Dugard Didn't Flee." Livescience.Com, 31 Aug. 2009, www.livescience .com/7862-bonding-captor-jaycee-dugard-flee.html.

9. "A Revealing Experiment: Brown v. Board and 'The Doll Test.'" NAACP Legal Defense and Educational Fund, 4 Mar. 2019, www .naacpldf.org/ldf-celebrates-60th-anniversary-brown-v-board -education/significance-doll-test. This is the original study: Almosaed, Nora. "Violence Against Women: A Cross-cultural Perspective." *Journal of Muslim Minority Affairs,* vol. 24, no. 1, 2004, pp. 67–88, doi:10.1080/1360200042000212124. Additional studies are cited in the World Health Organization's 2009 report titled *Changing Cultural and Social Norms That Support Violence.*

10. See, for instance: Amoakohene, M. "Violence Against Women in Ghana: A Look at Women's Perceptions and Review of Policy and Social Responses." *Social Science & Medicine,* vol. 59, no. 11, 2004, doi:10.1016/s0277-9536(04)00163-7.

11. Westcott, Kathryn. "What Is Stockholm Syndrome?" BBC News, 22 Aug. 2013, www.bbc.com/news/magazine-22447726.

CHAPTER 6: COSTLY SIGNALING AND AESTHETICS

1. See Chapter 13 of Darwin's *The Descent of Man,* for instance.

2. Although we focus on Zahavi's costly signaling explanation for long, ostentatious tails, there are others, most prominently that of runaway selection, first suggested by Darwin and expanded upon by Fisher; for a more recent take, see Richard Prum's *The Evolution of Beauty* (New York: Anchor, 2017). In runaway selection, once peahens develop a preference for long tails (perhaps, at first, because those tails have some function, or perhaps randomly), any female who mutates away from such a preference will fair worse not because it means she will mate with less fit peacocks, but because it means that her mates are more likely to have short tails, and hence so will her sons, leading them to have fewer mating opportunities given the prevailing preferences.

3. Evans, Matthew R., and B. J. Hatchwell. "An Experimental Study of Male Adornment in the Scarlet-Tufted Malachite Sunbird: I. The Role of Pectoral Tufts in Territorial Defence." *Behavioral Ecology and Sociobiology*, vol. 29, no. 6, 1992, doi:10.1007/bf00170171.

4. The first study we know of to use this tail-tampering technique is: Andersson, Malte. "Female Choice Selects for Extreme Tail Length in a Widowbird." *Nature*, vol. 299, no. 5886, 1982, pp. 818–820, doi:10.1038/299818a0.

5. Readers interested in conspicuous consumption might want to check out Robert H. Frank's *Luxury Fever* (Princeton, NJ: Princeton

University Press, 1999), which covers many more examples, as well as some policy implications.

6. We learned about this from Ken Albala's course, "Food: A Cultural Culinary History" (The Great Courses, www.thegreatcourses .com/courses/food-a-cultural-culinary-history). See also Maanvi Singh's "How Snobbery Helped Take the Spice Out of European Cooking" NPR, 26 Mar., 2015.

7. Shephard, Wade. "Why Chinese Men Grow Long Fingernails." *Vagabond Journey* (blog), 14 May 2013, www.vagabondjourney.com /why-chinese-men-grow-long-fingernails.

8. This application of the costly signaling model evolved out of discussions with Uri Gneezy.

9. *Skin Lightening Products Market Size, Share & Trends Analysis Report by Product, by Nature, by Region, and Segment Forecasts, 2019–2025.* Grand View Research, Aug. 2019, www.grandview research.com/industry-analysis/skin-lightening-products-market.

10. Schube, Sam, and Yang-Yi Goh. "The Best White Dress Shirts Are the Foundation to Any Stylish Guy's Wardrobe." *GQ*, 19 Sept. 2019, www.gq.com/story/the-best-white-dress-shirts.

11. Wagner, John A., and Susan Walters Schmid. *Encyclopedia of Tudor England.* Vol. 1, A–D, p. 277. Santa Barbara, CA: ABC-CLIO, 2011.

12. In fact, Paul Bloom's *How Pleasure Works* (New York: W. W. Norton, 2010) is filled with all sorts of quirky features of our sense of aesthetics. We highly recommend it. Geoffrey Miller's *The Mating Mind* (New York: Anchor, 2000) covers many other aspects of aesthetics that may be shaped by costly signaling, although that book's focus seems to be primarily about signaling to potential mates. Steven Pinker has another classic take on aesthetics, the proverbial "auditory cheesecake" originally discussed in *How the Mind Works*. That story has been expanded upon rather nicely by the work of V. S. Ramachandran, for example, in his paper: Ramachandran, V. S., and William Hirstein. "The Science of Art: A Neurological Theory of Aesthetic Experience." *Journal of Consciousness Studies*, vol. 6, no. 6–7, 1999, pp. 15–51.

13. The scene appears in season 1, episode 7.

14. We've borrowed this example from the informative folks at Wine Folly, authors of two well-known wine books and keepers of the world's largest wine database. Keeling, Phil. "50 of the Most

Eye-Rolling Wine Snob Moments." *Wine Folly*, 9 June 2020, winefolly.com/lifestyle/50-of-the-most-eye-rolling-wine-snob -moments.

15. This *Vox* video walks through the evolution of complex rhyming schemes, and includes some great examples and explanations: https://youtu.be/QWveXdj6oZU.

16. Mindel, Nissan. "Laws of the Morning Routine." Chabad.org. www.chabad.org/library/article_cdo/aid/111217/jewish/Laws-of-the -Morning-Routine.htm.

17. We highly recommend Ara Norenzayan's book *Big Gods* (Princeton, NJ: Princeton University Press, 2013) as well as Edward Slingerland and Azim Shariff's online course "The Science of Religion" (www.edx.org/course/the-science-of-religion). These include reviews of the evidence cited herein, plus much more cool stuff on the topic.

18. There are alternative explanations to Sosis's. Joe Henrich, for instance, has suggested that if people infer that anyone willing to engage in extreme religious rituals must have a good reason for doing so, then these rituals and the associated beliefs can spread, even if they don't serve a signaling function—or any other function, for that matter. See: Henrich, Joseph. "The Evolution of Costly Displays, Cooperation and Religion: Credibility Enhancing Displays and Their Implications for Cultural Evolution." *Evolution and Human Behavior* vol. 30, no. 4, 2009, pp. 244–260.

19. This argument as well as some of the upcoming evidence is first summarized in Sosis, Richard. "Why Aren't We All Hutterites?" *Human Nature*, vol. 14, no. 2, 2003, pp. 91–127, doi:10.1007/s12110 -003-1000-6.

20. See: Sosis, Richard, and Eric R. Bressler. "Cooperation and Commune Longevity: A Test of the Costly Signaling Theory of Religion." *Cross-Cultural Research*, vol. 37, no. 2, 2003, pp. 211–39, doi: 10.1177/1069397103037002003; and Sosis, Richard. "The Adaptive Value of Religious Ritual." *American Scientist*, vol. 92, no. 2, 2004, p. 166, doi:10.1511/2004.46.928.

21. See Chapter 20 in Stark, Rodney, *The Triumph of Christianity*, New York: HarperOne, 2011.

22. Sosis, R., Howard C. Kress, and James S. Boster. "Scars for War: Evaluating Alternative Signaling Explanations for Cross-Cultural Variance in Ritual Costs." *Evolution and Human Behavior*,

vol. 28, no. 4, 2007, pp. 234–247, doi:10.1016/j.evolhumbehav.2007
.02.007.

23. Soler, Montserrat. "Costly Signaling, Ritual and Cooperation:
Evidence from Candomblé, an Afro-Brazilian Religion." *Evolution
and Human Behavior*, vol. 33, no. 4, 2012, pp. 346–356, doi:10.1016/j
.evolhumbehav.2011.11.004.

CHAPTER 7: BURIED SIGNALS AND MODESTY

1. This quote is from the Greek philosopher Diogenes.

2. Lau, Melody. "Justin Bieber Gives Singer Carly Rae Jepsen a
Boost." *Rolling Stone*, 12 Mar. 2012.

3. "Grigory Sokolov: Biography." Deutsche Grammophon, Mar.
2020, www.deutschegrammophon.com/en/artists/grigory-sokolov
/biography.

4. Cooper, Michael. "Lang Lang Is Back: A Piano Superstar Grows
Up." *New York Times,* 26 July 2019, www.nytimes.com/2019/07/24
/arts/music/lang-lang-piano.html.

5. Qiu, Jane. "Rothko's Methods Revealed." *Nature*, vol. 456,
no. 7221, 2008, p. 447, doi:10.1038/456447a.

6. This chapter is based on a model developed with Christian
Hilbe and Martin Nowak: Hoffman, Moshe, Christian Hilbe, and
Martin A. Nowak. "The Signal-Burying Game Can Explain Why We
Obscure Positive Traits and Good Deeds." *Nature Human Behaviour*,
vol. 2, no. 6, 2018, pp. 397–404. It's also closely related to other
models, like those on countersignaling; see, for instance, Feltovich,
Nick, Richmond Harbaugh, and Ted To. "Too Cool for School? Sig-
nalling and Countersignalling." *RAND Journal of Economics*, 2002,
pp. 630–649.

7. Of course, modesty is sometimes motivated by a desire to avoid
social sanctions. This is common in egalitarian hunter gatherers. See,
for instance: R. B. Lee and I. DeVore, eds. *Kalahari Hunter-Gatherers:
Studies of the !Kung San and Their Neighbors.* Cambridge, MA: Harvard University Press, 1976. However, this explanation for modesty
leaves unexplained why we draw positive inferences about someone's
traits when we learn they have been modest.

8. Flegenheimer, Matt. "Thomas Kinkade, Painter for the Masses,
Dies at 54." *New York Times*, 8 Apr. 2012, www.nytimes.com
/2012/04/08/arts/design/thomas-kinkade-artist-to-mass-market-dies
-at-54.html.

CHAPTER 8: EVIDENCE GAMES AND SPIN

1. LaFata, Alexia, and Corinne Sullivan. "What Are the Best Tinder Bios to Get Laid? Here Are 4 Tips." *Elite Daily*, 8 June 2021, www.elitedaily.com/dating/best-tinder-bios-to-get-laid.

2. "How to Write an Effective Resume." The Balance Careers, www.thebalancecareers.com/job-resumes-4161923. Accessed 27 Aug. 2021.

3. Athey, Amber. "MSNBC reporter Gadi Schwartz busts his own network's narrative about the caravan: 'From what we've seen, the majority are actually men and some of these men have not articulated that need for asylum.'" Twitter, 26 Nov. 2018, twitter.com/amber_athey/status/1067163239853760512. In the United States, there are three cable TV channels devoted almost entirely to news: CNN, MSNBC, and Fox News. Politics features heavily in the programming of all three channels, and it's no secret that the channels are politically biased: CNN and MSNBC are both left of center, whereas Fox News is famously right wing.

4. Peters, Justin. "Fox News Is the Tarp on the MAGA Van." *Slate*, 27 Oct. 2018, slate.com/news-and-politics/2018/10/cesar-sayoc-fox-news-trump-fanaticism.html.

5. Smart, Charlie. "The Differences in How CNN, MSNBC, and FOX Cover the News." The Pudding, pudding.cool/2018/01/chyrons. Accessed 27 Aug. 2021.

6. Nicas, Jack. "Apple Reports Declining Profits and Stagnant Growth, Again." *New York Times*, 31 July 2019, www.nytimes.com/2019/07/30/technology/apple-earnings-iphone.html.

7. Baker, Peter. "Christine Blasey Ford's Credibility Under New Attack by Senate Republicans." *New York Times*, 3 Oct. 2018, www.nytimes.com/2018/10/03/us/politics/blasey-ford-republicans-kavanaugh.html.

8. Bertrand, Natasha. "FBI Probe of Brett Kavanaugh Limited by Trump White House." *Atlantic*, 24 Oct. 2018, www.theatlantic.com/politics/archive/2018/10/fbi-probe-brett-kavanaugh-limited-trump-white-house/572236.

9. Wikipedia contributors. "Five-Paragraph Essay." Wikipedia, 21 Apr. 2021, en.wikipedia.org/wiki/Five-paragraph_essay.

10. The Climate Reality Project. *The 12 Questions Every Climate Activist Hears and What to Say* (pamphlet). 2019, p. 4.

11. Watson, Kathryn. "Trump Approval Poll Offers No Negative Options, Asks about Media Coverage of Trump's Approval Rating." CBS News, 30 Dec. 2017, www.cbsnews.com/news/trump-approval -poll-offers-no-negative-options.

12. For more examples and causes of problematic research practices, see: Ritchie, Stuart. *Science Fictions: How Fraud, Bias, Negligence, and Hype Undermine the Search for Truth.* New York: Metropolitan Books, 2020.

13. By the way, it doesn't yet matter whether the sender knows the state. Later, we'll assume he does not.

14. Here're a few more details on how the calculation is done. Bayes' rule says: Probability (state is high and evidence is observed) = Probability (state is high and evidence is observed) / Probability (evidence is observed). Probability (state is high and evidence is observed) is just pq^h. Probability (evidence is observed) is the probability of obtaining evidence in both states, $(pq^h + (1 - p)q^l)$. Probability (state is high and evidence is observed) is therefore $pq^h / (pq^h + (1 - p)q^l)$. If we plug in the parameters $p = .3$, $q^h = 6\%$, and $q^l = .1\%$, we get $.3*6\% / (.3*6\% + .7*.1\%) = 96.25\%$.

15. Say, when $q^h - q^l$ is greater than .75.

16. Say, if and only if $p < .1$.

17. We also need to assume $0 < p < 1$ and $q^h \neq q^l$; though, in these cases the problem would become rather uninteresting, anyway.

18. Technically, we also need to assume that if the receiver were to observe evidence that she didn't expect the sender to show her, she would still update her beliefs based on this evidence.

19. This posterior is equal to the probability the evidence exists and is found in the high state, $pq^h f_{max}$, divided by the probability that evidence exists and is found in either state, $(pq^h + (1 - p)q^l)f_{max}$. Since the probability the evidence is found is the same in both states, this term cancels.

20. This denominator for this posterior is equal to the probability that evidence doesn't exist, $(p(1 - q^h) + (1 - p)(1 - q^l))$, plus the probability that it does exist but wasn't found even though the sender searched maximally, $(1 - f_{max})(pq^h + (1 - p)q^l)$. The numerator is: $p(1 - q^h) + pq^h(1 - f_{max})$.

21. Suppose the sender gets a payoff of $k > 0$ for each additional percentage point that the receiver's posterior ends up assigning to the high state. The condition is then $c < k (\Phi_{max} - \Phi_{min}) (\mu^1 - \mu^0)$, where $\Phi_{max} = pq^h f_{max} + (1 - p)q^l f_{max}$ is the probability of obtaining evidence when

searching maximally, $\Phi_{min} = pq^b f_{min} + (1 - p)q^l f_{min}$ is the probability of obtaining evidence when searching minimally, $\mu^1 = pq^b f_{max} / (pq^b f_{max} + (1 - p)q^l f_{max})$ is the receiver's posterior when she sees evidence and expects the sender to have searched maximally, and $\mu^0 = [p(q^b(1 - f_{max}) + (1 - q^b))] / [p(q^b(1 - f_{max}) + (1 - q^b)) + (1 - p)(q^l(1 - f_{max}) + 1 - q^l))]$ is the receiver's posterior when she doesn't see evidence and expects the sender to have searched maximally

22. We presume that the sender makes his choice without knowing the actual state.

23. Milgrom, Paul. "What the Seller Won't Tell You: Persuasion and Disclosure in Markets." *Journal of Economic Perspectives*, vol. 22, no. 2, 2008, pp. 115–131, doi:10.1257/jep.22.2.115.

24. OkCupid. "The Big Lies People Tell In Online Dating" (blog). 10 Aug. 2021, theblog.okcupid.com/the-big-lies-people-tell-in-online -dating-a9e3990d6ae2.

CHAPTER 9: MOTIVATED REASONING

1. See Hippel, William von, and Robert Trivers. "The Evolution and Psychology of Self-Deception." *Behavioral and Brain Sciences*, vol. 34, no. 1, 2011, pp. 1–16, doi:10.1017/s0140525x10001354; Trivers, Robert. *The Folly of Fools: The Logic of Deceit and Self-Deception in Human Life.* New York: Basic Books, 2014; and Kurzban, Robert. *Why Everyone (Else) Is a Hypocrite: Evolution and the Modular Mind.* New York: Basic Books, 2012. Or, for a more recent review, see: Williams, Daniel. "Socially Adaptive Belief." *Mind & Language*, vol. 36, no. 3, 2020, pp. 333–354, doi:10.1111/mila.12294.

2. For another classic example, see: Weinstein, Neil D. "Unrealistic Optimism About Future Life Events." *Journal of Personality and Social Psychology*, vol. 39, no. 5, 1980, p. 806. Weinstein finds that students predict that they're more likely than average to have good things happen to them, and less likely than average to have bad things happen to them.

3. Eil, David, and Justin M. Rao. "The Good News–Bad News Effect: Asymmetric Processing of Objective Information About Yourself." *American Economic Journal: Microeconomics*, vol. 3, no. 2, 2011, pp. 114–138, doi:10.1257/mic.3.2.114.

4. Gilbert, Daniel. "I'm O.K., You're Biased." *New York Times*, 16 Apr. 2006, www.nytimes.com/2006/04/16/opinion/im-ok-youre -biased.html.

5. Ditto, Peter H., and David F. Lopez. "Motivated Skepticism: Use of Differential Decision Criteria for Preferred and Nonpreferred Conclusions." *Journal of Personality and Social Psychology*, vol. 63, no. 4, 1992, pp. 568–584, doi:10.1037/0022-3514.63.4.568.

6. Lord, Charles G., Lee Ross, and Mark Lepper. "Biased Assimilation and Attitude Polarization: The Effects of Prior Theories on Subsequently Considered Evidence." *Journal of Personality and Social Psychology*, vol. 37, no. 11, 1979, pp. 2098–2109, doi:10.1037/0022-3514.37.11.2098.

7. Brooks, David. "How We Destroy Lives Today." *New York Times*, 22 Jan. 2019, www.nytimes.com/2019/01/21/opinion/covington-march-for-life.html.

8. Bloomberg Wire. "Rex Tillerson to Oil Industry: Not Sure Humans Can Do Anything to Battle Climate Change." *Dallas News*, 4 Feb. 2020, www.dallasnews.com/business/energy/2020/02/04/rex-tillerson-to-oil-industry-not-sure-humans-can-do-anything-to-battle-climate-change.

9. Weber, Harrison. "The Curious Case of Steve Jobs' Reality Distortion Field." VentureBeat, 24 Mar. 2015, venturebeat.com/2015/03/24/the-curious-case-of-steve-jobs-reality-distortion-field.

10. Babcock, Linda, George Loewenstein, Samuel Issacharoff, and Colin Camerer. "Biased Judgments of Fairness in Bargaining." *American Economic Review*, vol. 85, no. 5, 1995, 1337–1343.

11. Schwardmann, Peter, Egon Tripodi, and Joël J. Van der Weele. "Self-Persuasion: Evidence from Field Experiments at Two International Debating Competitions." SSRN, CESifo Working Paper No. 7946, 27 Nov. 2019.

12. In fact, Trump was ahead of his peers. In 2009, only 50 percent of Trump's fellow Democrats thought global warming was man-made. See: Kohut, Andrew, Carroll Doherty, Michael Dimock, and Scott Keeter. "Fewer Americans See Solid Evidence of Global Warming" (news release). Pew Research Center for the People & the Press, Washington, DC, 22 Oct. 2009.

13. Anthes, Emily. "C.D.C Studies Say Young Adults Are Less Likely to Get Vaccinated." *New York Times*, 21 June 2021, www.nytimes.com/2021/06/21/health/vaccination-young-adults.html.

14. Zimmermann, Florian. "The Dynamics of Motivated Beliefs." *American Economic Review*, vol. 110, no. 2, 2020, pp. 337–361.

15. Schwardmann, Peter, and Joel Van der Weele. "Deception and Self-Deception." *Nature Human Behaviour*, vol. 3, no. 10, 2019, pp. 1055–1061.

16. Kunda, Ziva. "The Case for Motivated Reasoning." *Psychological Bulletin*, vol. 108, no. 3, 1990, pp. 480–498, doi:10.1037/0033-2909.108.3.480.

17. Thaler, Michael. "Do People Engage in Motivated Reasoning to Think the World Is a Good Place for Others?" Cornell University, 2 Dec. 2020, arXiv:2012.01548.

18. See, for instance, work by our colleagues Dave Rand and Gord Pennycook, which shows that people's tendency to share fake news is largely driven by inattention. Pennycook, Gordon, and David G. Rand. "Lazy, Not Biased: Susceptibility to Partisan Fake News Is Better Explained by Lack of Reasoning Than by Motivated Reasoning." *Cognition*, vol. 188, 2019, pp. 39–50.

19. Thaler, Michael. "The 'Fake News' Effect: Experimentally Identifying Motivated Reasoning Using Trust in News." 22 July 2021, last updated 18 Aug. 2021, SSRN, https://ssrn.com/abstract=3717381.

CHAPTER 10: THE REPEATED PRISONER'S DILEMMA AND ALTRUISM

1. Some sources that may prove useful for those who want to take a deeper dive: Weibull, Jörgen. *Evolutionary Game Theory*. Cambridge, MA: MIT Press, 1995; Hofbauer, Josef, and Karl Sigmund. *Evolutionary Games and Population Dynamics*. First ed. Cambridge, UK: Cambridge University Press, 1998; Nowak, Martin. *Evolutionary Dynamics: Exploring the Equations of Life*. First ed. Cambridge, MA: Belknap Press, 2006; and Fudenberg, Drew, and David Levine. *The Theory of Learning in Games (Economic Learning and Social Evolution)*. Cambridge, MA: MIT Press, 1998.

2. Wilkinson, Gerald S. "Reciprocal Altruism in Bats and Other Mammals." *Ethology and Sociobiology*, vol. 9, no. 2–4, 1988, pp. 85–100, doi:10.1016/0162-3095(88)90015-5; and Carter, Gerald G., and Gerald S. Wilkinson. "Food Sharing in Vampire Bats: Reciprocal Help Predicts Donations More than Relatedness or Harassment." *Proceedings of the Royal Society B: Biological Sciences*, vol. 280, no. 1753, 2013, p. 20122573, doi:10.1098/rspb.2012.2573.

CHAPTER 11: NORM ENFORCEMENT

1. This model is adapted from: Panchanathan, Karthik, and Robert Boyd. "Indirect Reciprocity Can Stabilize Cooperation without the Second-Order Free Rider Problem." *Nature*, vol. 432, no. 7016, 2004, pp. 499–502, doi:10.1038/nature02978. For a review of related models of indirect reciprocity, see: Okada, Isamu. "A Review of Theoretical Studies on Indirect Reciprocity." *Games*, vol. 11, no. 3, 2020, p. 27, doi:10.3390/g11030027. Many of the insights in this chapter are summarized in: Boyd, Robert. *A Different Kind of Animal: How Culture Transformed Our Species.* Princeton, NJ: Princeton University Press, 2017. For a closely related perspective and many more examples, see: Bicchieri, Cristina. *Norms in the Wild: How to Diagnose, Measure, and Change Social Norms.* Oxford, UK: Oxford University Press, 2016.

Although we focus on the role that norm enforcement and repeated games more generally play in shaping our altruistic intuitions, other ultimate mechanisms also play a role and offer insight. For instance, another important mechanism seems to be signaling, sometimes referred to as competitive altruism because it can lead to an arms race when individuals try to outshine one another. The key idea is that giving is a costly signal: to help others, one must have resources to expend and more of a willingness to do so (which might, say, come from a need to be trusted). This literature is nicely summarized in Chapter 12 of Nichola Raihani's *The Social Instinct: How Cooperation Shaped the World* (New York: St. Martin's Press, 2021), which we recommend. An example of the kinds of insights that this mechanism would explain include the ratcheting effect illustrated in N. J. Raihani, and Smith, S. "Competitive Helping in Online Giving." *Current Biology*, vol. 25, 2015, pp. 1183–1186. In their study, men gave substantially more on a crowdfunding platform when they saw that another man had given a large gift (two standard deviations above the mean), and this effect was particularly pronounced—leading to four times the normal donations—when the fundraiser was an attractive female. Ratcheting of this kind seems to have been on frequent display in ancient Rome, where the Roman elite competed with one another for dignitas through military feats and public works like theaters and aqueducts.

A separate mechanism that's often discussed is "assortment," which refers to the fact that cooperators are liable to find themselves in the company of other cooperators. This can happen, for instance, because cultural and biological traits tend to spread locally, or because cooperators self-segregate. See, for instance, Ingela Alger and Jörgen W. Weibull's "Homo Moralis—Preference Evolution Under Incomplete Information and Assortative Matching." *Econometrica*, vol. 81, no. 6, 2013, pp. 2269–2302. Or Matthijs Van Veelen, Julián García, David Rand, and Martin A. Nowak's "Direct Reciprocity in Structured Populations." *Proceedings of the National Academy of Sciences*, vol. 109, no. 25, 2012, pp. 9929–9934. Many of the features of altruism and norms summarized in this chapter and the subsequent ones are hard to reconcile with these alternative mechanisms, but that doesn't mean that altruism and norms aren't also often driven by these alternative mechanisms.

2. Technically, the benefit to the group must be greater than C, but less than n^*C.

3. Fehr, Ernst, and Urs Fischbacher. "Third-Party Punishment and Social Norms." *Evolution and Human Behavior*, vol. 25, no. 2, 2004, pp. 63–87, doi:10.1016/s1090-5138(04)00005-4; and Fehr, Ernst, Urs Fischbacher, and Simon Gächter. "Strong Reciprocity, Human Cooperation, and the Enforcement of Social Norms." *Human Nature*, vol. 13, no. 1, 2002, pp. 1–25, doi:10.1007/s12110-002-1012-7.

4. Fehr, Ernst, and Simon Gächter. "Altruistic Punishment in Humans." *Nature*, vol. 415, no. 6868, 2002, pp. 137–140, doi:10.1038/415137a.

5. Henrich, Joseph, et al. "Costly Punishment Across Human Societies." *Science*, vol. 312, no. 5781, 2006, pp. 1767–1770.

6. We don't have a published citation for this; we just know he found these from discussions with him.

7. Webb, Clive. "Jewish Merchants and Black Customers in the Age of Jim Crow." *Southern Jewish History*, vol. 2, 1999, pp. 55–80.

8. Kurzban, Robert, Peter DeScioli, and Erin O'Brien. "Audience Effects on Moralistic Punishment." *Evolution and Human Behavior*, vol. 28, no. 2, 2007, pp. 75–84, doi:10.1016/j.evolhumbehav.2006.06.001.

9. Jordan, Jillian J., and Nour Kteily. "Reputation Drives Morally Questionable Punishment." Harvard Business School, Working Paper, December 2020.

10. Mathew, Sarah. "How the Second-Order Free Rider Problem Is Solved in a Small-Scale Society." *American Economic Review*, vol. 107, no. 5, 2017, pp. 578–581, doi:10.1257/aer.p20171090.

11. You might wonder why Robinson's teammates, some of whom were, themselves, quite racist, swallowed their own inclinations and tolerated the abuse thrust upon them. It's because the Dodgers manager threatened, "I do not care if the guy is yellow or black, or if he has stripes like a fuckin' zebra. I'm the manager of this team, and I say he plays. What's more, I say he can make us all rich. And if any of you cannot use the money, I will see that you are all traded." For its part, the National Baseball League came down hard on any player or team that threatened to strike. The league's president is quoted as telling the players, "You will find that the friends that you think you have in the press box will not support you, that you will be outcasts. I do not care if half the league strikes. Those who do it will encounter quick retribution. All will be suspended and I don't care if it wrecks the National League for five years. This is the United States of America and one citizen has as much right to play as another."

12. Wikipedia Contributors. "List of Excommunicable Offences in the Catholic Church." Wikipedia, 9 Feb. 2021, en.wikipedia.org/wiki /List_of_excommunicable_offences_in_the_Catholic_Church.

13. Hamlin, J. K., Karen Wynn, Paul Bloom, and Neha Mahajan. "How Infants and Toddlers React to Antisocial Others." *Proceedings of the National Academy of Sciences*, vol. 108, no. 50, 2011, pp. 19931–19936, doi:10.1073/pnas.1110306108.

14. Franzen, Axel, and Sonja Pointner. "Anonymity in the Dictator Game Revisited." *Journal of Economic Behavior & Organization*, vol. 81, no. 1, 2012, pp. 74–81, doi:10.1016/j.jebo.2011.09.005.

15. List, John A., Robert P. Berrens, Alok K. Bohara, and Joe Kerkvliet. "Examining the Role of Social Isolation on Stated Preferences." *American Economic Review*, vol. 94, no. 3, 2004, pp. 741–752, doi:10.1257/0002828041464614.

16. Bandiera, Oriana, Iwan Barankay, and Imran Rasul. "Social Preferences and the Response to Incentives: Evidence from Personnel Data." *Quarterly Journal of Economics*, vol. 120, no. 3, 2005, pp. 917–962, doi:10.1162/003355305774268192.

17. Yoeli, E., M. Hoffman, D. G. Rand, and Martin A. Nowak. "Powering Up with Indirect Reciprocity in a Large-Scale Field Experiment." *Proceedings of the National Academy of Sciences*, vol. 110, supplement 2, 2013, pp. 10424–10429, doi:10.1073/pnas.1301210110.

18. For a review, see: Kraft-Todd, Gordon, Erez Yoeli, Syon Bhanot, and David Rand. "Promoting Cooperation in the Field." *Current*

Opinion in Behavioral Sciences, vol. 3, June 2015, pp. 96–101, doi:10.1016/j.cobeha.2015.02.006.

19. List, John A. "On the Interpretation of Giving in Dictator Games." *Journal of Political Economy*, vol. 115, no. 3, 2007, pp. 482–493, doi:10.1086/519249.

20. Liberman, Varda, Steven M. Samuels, and Lee Ross. "The Name of the Game: Predictive Power of Reputations versus Situational Labels in Determining Prisoner's Dilemma Game Moves." *Personality and Social Psychology Bulletin*, vol. 30, no. 9, 2004, pp. 1175–1185.

21. Capraro, Valerio, and Andrea Vanzo. "The Power of Moral Words: Loaded Language Generates Framing Effects in the Extreme Dictator Game." *Judgment and Decision Making*, vol. 14, no. 3, 2019, pp. 309–317.

22. Goldstein, Noah J., Robert B. Cialdini, and Vladas Griskevicius. "A Room with a Viewpoint: Using Social Norms to Motivate Environmental Conservation in Hotels." *Journal of Consumer Research*, vol. 35, no. 3, 2008, pp. 472–482, doi:10.1086/586910.

23. See: Kraft-Todd, Gordon, Erez Yoeli, Syon Bhanot, and David Rand. "Promoting Cooperation in the Field." *Current Opinion in Behavioral Sciences*, vol. 3, 2015, pp. 96–101.

24. For more on the practical implications, check out this TEDx talk given by Erez but which we jointly authored: go.ted.com/erezyoeli. We also published this accompanying article explaining the underlying science behind our advice: Yoeli, Erez. "Is the Key to Successful Prosocial Nudges Reputation?" Behavioral Scientist, 31 July 2018, http://behavioralscientist.org/is-reputation-the-key-to-prosocial -nudges/. The papers on which this discussion is based are: Yoeli, Erez, Moshe Hoffman, David G. Rand, and Martin A. Nowak. "Powering Up with Indirect Reciprocity in a Large-Scale Field Experiment." *Proceedings of the National Academy of Sciences*, vol. 110, supplement 2, 2013, pp. 10424–10429; Rand, David G., Erez Yoeli, and Moshe Hoffman. "Harnessing Reciprocity to Promote Cooperation and the Provisioning of Public Goods." *Policy Insights from the Behavioral and Brain Sciences*, vol. 1, no. 1, 2014, pp. 263–269; and Yoeli, Erez, et al. "Digital Health Support in Treatment for Tuberculosis." *New England Journal of Medicine*, vol. 381, no. 10, 2019, pp. 986–987. The digital health intervention we reference was developed in collaboration with Jon Rathauser and the entire team at Keheala.

25. For yet more ways in which norms vary, see Michele Gelfand's *Rule Makers, Rule Breakers: Tight and Loose Cultures and the Secret Signals That Direct Our Lives.* New York: Scribner, 2019.

26. The quote is attributed to de Bucquoy (1744) and can be found in: Leeson, Peter T. "An-arrgh-chy: The Law and Economics of Pirate Organization." *Journal of Political Economy*, vol. 115, no. 6, 2007, pp. 1049–1094, doi:10.1086/526403.

27. For more on pirates, check out Peter Leeson's *The Invisible Hook: The Hidden Economics of Pirates.* Princeton, NJ: Princeton University Press, 2011.

28. Hengel, Brenda. "The Hit That Could Have Sunk Las Vegas." The Mob Museum, 25 June 2017, themobmuseum.org/blog/costello -hit-sunk-las-vegas.

29. Persio, Sofia Lotto. "Secret Courts Uncovered Where Mobsters Face Death for Breaking Mafia Code." *Newsweek*, 4 July 2017, www .newsweek.com/underground-mafia-courts-revealed-massive-bust -against-italian-ndrangheta-631650.

30. Al-Gharbi, Musa. "What Police Departments Do to Whistle-blowers." *Atlantic*, 1 July 2020, www.theatlantic.com/ideas/archive /2020/07/what-police-departments-do-whistle-blowers/613687.

31. Acheson, James. *The Lobster Gangs of Maine.* Amsterdam: Amsterdam University Press, 2012.

32. Ellickson, Robert C. "Of Coase and Cattle: Dispute Resolution Among Neighbors in Shasta County." *Stanford Law Review*, vol. 38, no. 3, 1986, p. 623. doi:10.2307/1228561.

33. The source for this section is Ad van Liempt's *Kopgeld* (Amsterdam: Balans, 2002), summarized in English in a number of articles that came out when it was published, like this one: Deutsch, Anthony. "Nazis Paid Bounty Hunters to Turn in Jews, Book Says." *Los Angeles Times*, 1 Dec. 2002, www.latimes.com/archives/la-xpm-2002-dec -01-adfg-bountyhunt1-story.html.

34. Jordan, Jillian J., Moshe Hoffman, Paul Bloom, and David G. Rand. "Third-Party Punishment as a Costly Signal of Trustworthiness." *Nature*, vol. 530, no. 7591, 2016, pp. 473–476, doi:10.1038 /nature16981.

35. Henrich, Joseph, and Michael Muthukrishna. "The Origins and Psychology of Human Cooperation." *Annual Review of Psychology*, vol. 72, no. 1, 2021, pp. 207–240, doi:10.1146/annurev-psych -081920-042106.

36. Where do all these norms and the institutions that help to enforce them come from? Why do they so often end up serving the group's interest, at least to the extent possible given the kinds of constraints on enforcement we discussed? To answer these questions, we'd need to consider additional mechanisms, beyond the stylized model we presented. One possibility is that cultural groups that develop more group-beneficial norms fare better, spreading their norms to other groups or getting more people to join their group. However, other possibilities abound. For instance, it could be that whoever is able to shape norms and institutions will sometimes have an interest in selecting group-beneficial ones. Historically, despots like Augustus and Genghis Khan have often encouraged norms that support strong property rights and trade and market activity, which benefits their subjects, while filling their coffers via taxes on agriculture output and trade. Some works that cover these possibilities include: Henrich, Joseph. "Cultural Group Selection, Coevolutionary Processes and Large-Scale Cooperation." *Journal of Economic Behavior & Organization*, vol. 53, no. 1, 2004, pp. 3–35; Richerson, Peter, et al. "Cultural Group Selection Plays an Essential Role in Explaining Human Cooperation: A Sketch of the Evidence." *Behavioral and Brain Sciences,* vol. 39, 2016; and Singh, Manvir, Richard Wrangham, and Luke Glowacki. "Self-Interest and the Design of Rules." *Human Nature*, vol. 28, no. 4, 2017, pp. 457–480.

CHAPTER 12: CATEGORICAL NORMS

1. This chapter is based on a paper we coauthored with Aygun Dalkiran and Martin Nowak: "Categorical Distinctions Facilitate Coordination" (SSRN, 18 Dec. 2020, available at https://ssrn.com /abstract=3751837). The model builds off of game theory models known as Global Games, which were first introduced by Hans Carlsson and Erik van Damme in "Global Games and Equilibrium Selection." *Econometrica*, vol. 61, no. 5, September 1993, pp. 989–1018. They were further developed by Stephen Morris and Hyun Song Shin in "Global Games: Theory and Applications." Chapter 3 in *Advances in Econometrics: Theory and Applications, Eighth World Congress*, 56–114. Edited by M. Dewatripont, L. Hansen, and S. Turnovsky. Cambridge: Cambridge University Press, 2003.

2. As in earlier chapters, when we talk about players' beliefs, we don't really mean conscious beliefs (that's something proximate that these models are meant to explain). We mean what an objective Bayesian observer would think.

3. We will be ignoring the instances where the signal falls within this small range of 0 and 1. Such edge cases don't really affect our results but do make the analysis more cumbersome.

4. Technically, so long as $(\mu(1 - \varepsilon)^2 + (1 - \mu)\varepsilon^2) / (\mu(1 - \varepsilon) + (1 - \mu)\varepsilon) \geq p$ and $(\mu\varepsilon^2 + (1 - \mu)(1 - \varepsilon)^2) / (\mu\varepsilon + (1 - \mu)(1 - \varepsilon)) \geq 1 - p$.

5. BBC News. "Why Has the Syrian War Lasted 10 Years?" BBC News, 12 Mar. 2021, www.bbc.com/news/world-middle-east-3580 6229.

6. See, for instance: McElreath, Richard, Robert Boyd, and Peter J. Richerson. "Shared Norms and the Evolution of Ethnic Markers." *Current Anthropology*, vol. 44, no. 1, no. 2003, pp. 122–130; Smedley, Audrey, and Brian Smedley. *Race in North America: Origin and Evolution of a Worldview.* Abingdon-on-Thames, UK: Routledge, 2018; and Moya, Cristina. *What Does It Mean for Humans to Be Groupish?* 2021.

7. Interestingly, throughout all this, both sides continued to avoid targeting civilians per se. This norm held up for about a year or so longer, but by mid-1942, the Allies were firebombing West German cities like Lubeck and Dresden with the explicit aim of terrorizing the civilian population.

8. Burum, Bethany, Martin A. Nowak, and Moshe Hoffman. "An Evolutionary Explanation for Ineffective Altruism." *Nature Human Behaviour*, vol. 4, no. 12, 2020, pp. 1245–1257, doi:10.1038/s41562-020-00950-4.

CHAPTER 13: HIGHER-ORDER BELIEFS

1. This chapter builds on the work of a number of researchers. Ariel Rubinstein highlighted that higher-order beliefs can matter for coordination in the now classic electronic mail game. Nobel laureate Robert Aumann first formalized the closely related concept of common knowledge. Dov Mondarer and Dov Samet formalized the notion of common *p*-beliefs in "Approximating Common Knowledge with Common Beliefs." *Games and Economic Behavior*, vol. 1, no. 2, June 1989, pp. 170–190. This is the formalism that we are using behind the scenes. Michael Suk-Young Chwe's *Rational Ritual: Culture,*

Coordination, and Common Knowledge (Princeton, NJ: Princeton University Press, 2003) explores social applications of higher-order beliefs. Steven Pinker's *The Stuff of Thought* (New York: Viking, 2008) also covers many interesting applications and evidence, as does his subsequent work with students Kyle Thomas, James Lee, and Julia De Freitas. Peter DeScioli and Rob Kurzban had many of the insights covered in this chapter; see especially their paper "A Solution to the Mysteries of Morality." *Psychological Bulletin*, vol. 139, no. 2, July 2012. Last but definitely not least, the specific models presented in this chapter were developed jointly with Aygun Dalkiran.

2. See the very end of this video: www.youtube.com/watch?v =OxnWGaxtqwA.

3. The *Struma* disaster is summarized in detail in this Wikipedia page: Wikipedia Contributors. "*Struma* disaster." Wikipedia, 29 June 2021, en.wikipedia.org/wiki/Struma_disaster.

4. This puzzle has a long history among moral psychologists and has received interesting treatments from, for example, Mark Spranca, Fiery Cushman, Josh Greene, and Jonathan Baron. The explanation we ultimately build upon was introduced to us by Peter DeScioli and Robert Kurzban in "Mysteries of Morality." *Cognition*, vol. 112, no. 2, August 2009, pp. 281–299.

5. Rupar, Aaron. "Ivanka Trump's Viral G20 Video, Explained." *Vox*, 1 July 2019, www.vox.com/2019/7/1/20677253/ivanka-trump -g20-nepotism.

6. It is not crucial to assume that there are no false positives, but it makes it easier to see the role of second-order beliefs.

7. Snyder, Melvin L., Robert E. Kleck, Angelo Strenta, and Steven J. Mentzer. "Avoidance of the Handicapped: An Attributional Ambiguity Analysis." *Journal of Personality and Social Psychology* vol. 37, no. 12, 1979, p. 2297.

8. Schmitt, Eric. "Clinton's 'Sorry' to Pakistan Ends Barrier to NATO." *New York Times*, 5 July 2012, www.nytimes.com/2012/07/04 /world/asia/pakistan-opens-afghan-routes-to-nato-after-us-apology .html.

9. This phrase is borrowed from John L. Austin, who coined it in his book *How to Do Things with Words*. Second ed. Oxford, UK: Oxford University Press, 1975.

10. Fiske, Alan P. "The Four Elementary Forms of Sociality: Framework for a Unified Theory of Social Relations." *Psychological Review*, vol. 99, no. 4, 1992, pp. 689–723, doi:10.1037/0033-295x.99.4.689.

11. For interesting alternative explanations to the one we're about to present, see the work of Josh Greene and Fiery Cushman. For instance: Greene, Joshua. *Moral Tribes: Emotion, Reason, and the Gap Between Us and Them*. London: Atlantic Books, 2021; and Cushman, Fiery. "Is Non-consequentialism a Feature or a Bug?" In *The Routledge Handbook of Philosophy of the Social Mind*, pp. 278–295. Abingdon-on-Thames, UK: Routledge, 2016.

12. Pistone, Joseph D. with Richard Woodley. *Donnie Brasco: My Undercover Life in the Mafia*. New York: Signet, 1989.

13. Kurzban, Robert, Peter DeScioli, and Daniel Fein. "Hamilton vs. Kant: Pitting Adaptations for Altruism against Adaptations for Moral Judgment." *Evolution and Human Behavior*, vol. 33, no. 4, 2012, pp. 323–333, doi:10.1016/j.evolhumbehav.2011.11.002.

14. Andreoni, James, Justin M. Rao, and Hannah Trachtman. "Avoiding the Ask: A Field Experiment on Altruism, Empathy, and Charitable Giving." *Journal of Political Economy*, vol. 125, no. 3, 2017, pp. 625–653, doi:10.1086/691703.

15. Dana, Jason, Daylian M. Cain, and Robyn M. Dawes. "What You Don't Know Won't Hurt Me: Costly (but Quiet) Exit in Dictator Games." *Organizational Behavior and Human Decision Processes*, vol. 100, no. 2, 2006, pp. 193–201, doi:10.1016/j.obhdp.2005.10.001.

16. Dana, Jason, Roberto A. Weber, and Jason Xi Kuang. "Exploiting Moral Wiggle Room: Experiments Demonstrating an Illusory Preference for Fairness." *Economic Theory*, vol. 33, no. 1, 2006, pp. 67–80, doi:10.1007/s00199-006-0153-z.

17. Bava Metzia, 62a. Original text and translation can be found here: www.sefaria.org/Bava_Metzia.61a.4?lang=bi.

18. Hauser, Marc, et al. "A Dissociation Between Moral Judgments and Justifications." *Mind & Language*, vol. 22, no. 1, 2007, pp. 1–21, doi:10.1111/j.1468-0017.2006.00297.x.

CHAPTER 14: SUBGAME PERFECTION AND JUSTICE

1. For a summary of the feud, see Dean King's *The Feud: The Hatfields and McCoys; The True Story*. New York: Little, Brown, 2014.

2. Kingsley, Patrick, and Isabel Kershner. "After Raid on Aqsa Mosque, Rockets From Gaza and Israeli Airstrikes." *New York Times*, 19 May 2021, www.nytimes.com/2021/05/10/world/middle east/jerusalem-protests-aqsa-palestinians.html.

3. Al Jazeera. "Israel-Hamas Ceasefire Holds as UN Launches Gaza Aid Appeal." Gaza News, Al Jazeera, 24 May 2021, www .aljazeera.com/news/2021/5/23/israel-gaza-ceasefire-holds-as -un-launches-appeal-for-aid.

4. Mayo Clinic Staff. "Forgiveness: Letting Go of Grudges and Bitterness." Mayo Clinic, 13 Nov. 2020, www.mayoclinic.org/healthy -lifestyle/adult-health/in-depth/forgiveness/art-20047692.

5. Robert H. Frank's *Passions Within Reason: The Strategic Role of the Emotions* (New York: W. W. Norton, 1988) also offers an explanation for why we punish even when it might seem counterproductive. His argument rests on the ability to observably commit—a slightly different story from the one we present here. However, at some places, Frank also offers the interpretation that emotions like anger and love, as well as moral principles that cause people to overlook costs and benefits, are incentive compatible due to repeated interactions and reputations. This interpretation is consistent with the one we present in this chapter.

6. We've made the roles of player 1 and player 2 asymmetric just for convenience. In reality, each usually has the opportunity both to transgress and to punish the other. This would slightly change the math but not the main lessons of the chapter.

7. If you have a good memory, you might recall we made the same assumption in the norm enforcement game from Chapter 9 and might also notice the contrast with the repeated prisoner's dilemma of Chapter 8. There, the only way to punish is to withhold cooperation and doing so actually benefits the player doing the punishing because he or she gets to avoid paying the cost of cooperation for the duration of the punishment.

8. See, for instance: "Neville Chamberlain: Heroic Peacemaker or Pathetic Pushover?" Sky History TV Channel, www.history.co.uk /article/neville-chamberlain-heroic-peacemaker-or-pathetic-pushover. Accessed 27 Aug. 2021.

9. For another interesting take on revenge and forgiveness, check out Michael Mccullough's *Beyond Revenge* (San Francisco: Jossey-Bass, 2008).

10. Shelton, Jacob. "The Sausage Duel: When Two Politicians Almost Faced Off Using Poisoned Meat." History Daily, 11 Mar. 2021, historydaily.org/sausage-duel-facts-stories-trivia.

11. "Abraham Lincoln's Duel." American Battlefield Trust, 25 Mar. 2021, www.battlefields.org/learn/articles/abraham-lincolns-duel.

12. Wikipedia contributors. "Duel." *Wikipedia*, 19 Aug. 2021, en.wikipedia.org/wiki/Duel.

13. Wells, C. A. "The End of the Affair: Anti-dueling Laws and Social Norms in Antebellum America." *Vanderbilt Law Review*, vol. 54, 2001, p. 1805.

14. Nagel, Thomas. "Moral Luck." Chap. 3 in *Mortal Questions*, pp. 24–38. New York: Cambridge University Press, 1979.

15. Cushman, Fiery, Anna Dreber, Ying Wang, and Jay Costa. "Accidental Outcomes Guide Punishment in a 'Trembling Hand' Game." *PLoS ONE*, vol. 4, no. 8, 2009, p. e6699, doi:10.1371/journal.pone .0006699.

16. Luckhurst, Toby. "The DMZ 'Gardening Job' That Almost Sparked a War." BBC News, 21 Aug. 2019, www.bbc.com/news /world-asia-49394758.

17. Ash, Elliott, Daniel L. Chen, and Suresh Naidu. "Ideas Have Consequences: The Impact of Law and Economics on American Justice." *Center for Law & Economics Working Paper Series*, vol. 4, 2019; and Drum, Kevin. "Here's How a Quiet Seminar Program Changed American Law." *Mother Jones*, 18 Oct. 2018, www.mother jones.com/kevin-drum/2018/10/heres-how-a-quiet-seminar-program -changed-american-law.

CHAPTER 15: THE HIDDEN ROLE OF PRIMARY REWARDS

1. Kozlowski, Joe. "Olympic Swimmer Katie Ledecky Is Worth $4 Million, but That's Nothing Compared to Her Uncle's $340 Million Fortune." Sportscasting, 23 July 2021, www.sportscasting.com /olympic-swimmer-katie-ledecky-is-worth-4-million-thats-nothing -uncle-340-million-fortune.

2. This quote is from *Itzhak*, the documentary we referenced in the introduction.

3. This is a reference to a classic paper by Sherwin Rosen called "The Economics of Superstars." *American Economic Review*, vol. 71, no. 5, 1981, pp. 845–858. For more examples, and an accessible discussion of causes and consequences, see Robert H. Frank and Philip J. Cook's *The Winner-Take-All Society: Why the Few at the Top Get So Much More Than the Rest of Us* (New York: Penguin, 1995).

4. See Angela Duckworth's *Grit: The Power of Passion and Perseverance*. New York: Scribner, 2018.

5. Deci, Edward L. "The Effects of Contingent and Noncontingent Rewards and Controls on Intrinsic Motivation." *Organizational Behavior and Human Performance*, vol. 8, no. 2, 1972, pp. 217–229.

6. Benabou, Roland, and Jean Tirole. "Intrinsic and Extrinsic Motivation." *Review of Economic Studies*, vol. 70, no. 3, 2003, pp. 489–520, doi:10.1111/1467-937x.00253.

7. For a review, see: Maier, Steven F., and Martin E. Seligman. "Learned Helplessness: Theory and Evidence." *Journal of Experimental Psychology: General*, vol. 105, no. 1, 1976, p. 3.

INDEX

© Erez Yoeli

Moshe Hoffman is a research scientist at the Max Planck Institute for Evolutionary Biology and lecturer at Harvard's department of economics. His research focuses on using game theory, models of learning and evolution, and experimental methods to decipher the motives that shape our social behavior, preferences, and ideologies.

© Rachel Tine

Erez Yoeli is a research scientist at MIT's Sloan School of Management, where he directs the Applied Cooperation Team (ACT). He is also a lecturer at Harvard's department of economics. His research focuses on altruism: understanding how it works and how to promote it.